INTERNATIONAL CENTRE FOR MECHANICAL SCIENCES

COURSES AND LECTURES - No. 95

JOSEF L. ZEMAN
TECHNICAL UNIVERSITY OF VIENNA

APPROXIMATE ANALYSIS OF STOCHASTIC PROCESSES IN MECHANICS

COURSE HELD AT THE DEPARTMENT
OF GENERAL MECHANICS
OCTOBER 1971

UDINE 1971

SPRINGER-VERLAG WIEN GMBH

Originally published by Springer-Verlag Wien-New York in 1972

ISBN 978-3-211-81131-3 ISBN 978-3-7091-2740-7 (eBook)
DOI 10.1007/978-3-7091-2740-7

PREFACE

Successful analysis of a physical system depends not only on the suitable selection of the mathematical methods but also on the proper choice of the model of the system. This model, obtained by isolating and idealizing the physical system and by discarding inessential details, should be sufficiently simple, to admit of mathematical solution, but, at the same time, should correspond to the physical system in some useful sense, to admit of theoretical conclusions based on some form of inductive reasoning. Of course, the correspondence between the behavior of the model and that of the physical system with its complex interaction with its surrounding can be approximate only.

Classical methods of analysis are based on models and inputs which are deterministic. There it is assumed that the inputs are known with complete certainty, and so are the properties of the materials, the geometric sizes and shapes. Of course, all the phenomena are deterministic then.

But perfect information is never true, physics and engineering are concerned with what can be measured, and measurements are always subject to inaccuracies. The interactions of the physical system with its surrounding, modelled by the inputs, are under the influence of many diverse effects, barely controlled, complexly interacting. There follows, that the most adequate models are frequently random ones with stochastic inputs. Fortunately in many cases deterministic models but with stochastic inputs suffice.

There is another reason for considering models with stochastic inputs. Frequently one is interested in the response of systems not only for several common fixed inputs but also for a whole spectrum of possible inputs that may arise under real conditions, and most easily this answer can be given by means of models with stochastic inputs.

This monograph, which is intended as textbook to my lectures given at the International

Center for Mechanical Sciences in October 1971, deals with the response of deterministic nonlinear memoryless or differential systems to stochastic perturbations. A relatively large part of this text is devoted to the Appendix, which consists of definitions and theorems required in the main text. This appendix is self-contained and followed by its own subject index. Three of its four sections are followed by lists of textbooks. These lists are by far not complete, listed are only those books I have consulted during preparation of the manuscript. A little more extensive is the list of literature given at the end of the main text.

In concluding the preface I take the opportunity to express my sincerest thanks to the authorities of CISM, especially to Prof. L. Sobrero and W. Olszak, for the invitation to present these lectures, and to my teacher Prof. H. Parkus, who initiated my interest in the applications of stochastic processes to problems of mechanics, and to whom I owe so much.

J. Leman

Udine, October, 1971.

Chapter 1
INTRODUCTION

In the Preface the importance of statistical investigations of mechanical systems has been indicated. It has been mentioned that for some problems the statistical investigation is the most adequate means, for others it is an auxiliary one supplementing the deterministic methods of investigation.

In the analysis of mechanical systems one has also to allow for the fact that all mechanical systems are non-linear - if at all then in a limited range only can linear relations describe mechanical systems with sufficient accuracy.

The study of the behavior of nonlinear systems (mathematical models) under the influence of stochastic perturbations thus is of vital interest for engineering investigations.

Exact solutions for the response of nonlinear systems to stochastic inputs are rare; in most of the problems encountered in engineering one is forced to use approximation methods. Especially attractive for engineering purposes are simple techniques, techniques that give quickly and with manageable amount of work estimates of (simple) statistical characteristics of system responses; and the result should be such that the influence of system parameters can be investigated easily.

Such simple techniques for investigating the response of nonlinear mechanical systems to stochastic perturbations are discussed in the following. Because of their importance in engineering only continuous-state continuous-time stochastic perturbations are considered and emphasis is put on moment characteristics, especially on moment functions of first and second order.

For simplicity only stochastic perturbations of equilibrium states of deterministic systems are considered. That means that such interesting problems as the investigation of the behavior of self-oscillating systems under influence of stochastic perturbations or of the response of nonlinear mechanical sytems to harmonic excitations with superposed stochastic ones are excluded. (Such problems, which are of special importance in control theory and the theory of radio communications and radar, are discussed for instance in [8, 10]). And it means that methods are excluded also which use random models of the original engineering systems.

Inasmuch as deviations of system states from an equilibrium state are of interest the latter is assumed to be a zero state and zero-state responses of the systems are investigated only.

Chapter 2

MEMORYLESS SYSTEMS

Nonlinear memoryless systems, the simplest form of nonlinear systems at all, are frequently encountered in engineering analysis, as models of real engineering systems of their own or as components of more complicated ones.

2.1. Exact Techniques

For a memoryless system the present values of the outputs depend solely on the present values of the inputs. Let functions g_i , $i = 1, \dots, n$ define the transformation of m stochastic inputs $U_1(t), \dots, U_m(t)$ into n outputs $Z_1(t), \dots, Z_n(t)$:

$$Z_i(t) = g_i\left[U_1(t), \dots, U_m(t)\right] , \quad i = 1, \dots, n . \qquad (2.1.1)$$

The random variables $Z_i(t_k), i = 1, \dots, n$, where t_k is an arbitrary moment of time, depend only on the random variables $U_j(t_k)$, $j = 1, \dots, m$. There follows that the problem of determining the statistical properties of the stochastic processes $Z_i(t)$ reduces to that of calculating the statistical properties of nonlinear functions of random variables, a problem discussed in the Appendix.

For instance, using the definition of the <u>probability distribution function</u>, one obtains

$$F_{Z_i(t)}(z_i) = P[Z_i(t) \leq z_i]$$

(2.1.2a)

$$= \int \ldots \int f_{U_1(t) \ldots U_m(t)}(u_1, \ldots, u_m) du_1 \ldots du_m \,,$$

where the integral has to be extended over the whole region for which

(2.1.2b) $$g_i(u_1, \ldots, u_m) \leq z_i \,, \quad i = 1, \ldots, n \,.$$

For continuous, one-time partially differentiable one-to-one functions g_i Eq. (B.49) is valid. For the single-input single-output case the <u>probability density function</u> follows as

(2.1.3) $$f_{Z(t)}(z) = f_{U(t)}(h(z)) \left| \frac{dh}{dz} \right| \,,$$

where $h(z)$ is the inverse of $g(u)$. This relation may also be used if the function g is not one-to-one but the inverse h is single-valued or breaks up into k single-valued branches $h_i(z)$:

(2.1.4) $$f_{Z(t)}(z) = \sum_{i=1}^{k} f_{U(t)}(h_i(z)) \left| \frac{dh_i}{dz} \right| \,.$$

Similarly Eq. (B.49) may be generalized.

<u>Gaussian input</u>:

For future use, some relations for statistical

characteristics of the output process $Z(t)=g(U(t))$ of a memoryless system whose single input $U(t)$ is Gaussian will be derived in the remainder of this section.

The two-dimensional probability density function of $U(t)$ may be represented in the form

$$f_{U(t_1)U(t_2)}(u_1,u_2) = \frac{1}{2\pi\sigma_1\sigma_2\sqrt{1-\varrho^2}} \cdot \exp\left[-\frac{\sigma_2^2(u_1-m_1)^2-2\sigma_1\sigma_2\varrho(u_1-m_1)(u_2-m_2)+\sigma_1^2(u_2-m_2)^2}{2\sigma_1^2\sigma_2^2(1-\varrho^2)}\right], \quad (2.1.5)$$

where the abbreviations

$$m_i = E\{U(t_i)\}, \quad i = 1,2,$$
$$\sigma_i^2 = Var\{U(t_i)\}, \quad i = 1,2, \quad (2.1.5a)$$

and

$$\varrho = Cov\{U(t_1),U(t_2)\}/\sigma_1\sigma_2$$

have been used. This probability density function (2.1.5) admits the expansion [5]

$$\frac{1}{\sigma_1\sigma_2}\phi\left(\frac{u_1-m_1}{\sigma_1}\right)\phi\left(\frac{u_2-m_2}{\sigma_2}\right)\sum_{n=0}^{\infty}\frac{1}{n!}\varrho^n H_n\left(\frac{u_1-m_1}{\sigma_1}\right)H_n\left(\frac{u_2-m_2}{\sigma_2}\right) \quad (2.1.6)$$

where $\phi(.)$ is the normal function defined by Eq. (B.20), and $H_n(.)$ is the nth Hermitian polynomial, defined by

(2.1.7) $$H_n(x) = (-1)^n \exp(x^2/2) \frac{d^n}{dx^n} \exp(-x^2/2) \ .$$

The Hermitian polynomials are orthogonal on $[-\infty,\infty]$ with weight $\exp(-x^2/2)$:

(2.1.7a) $$\int_{-\infty}^{\infty} H_n(x) H_m(x) \exp(-x^2/2) dx = \begin{cases} n!\sqrt{2\pi} & n = m \\ 0 & n \neq m \ . \end{cases}$$

There follows that

(2.1.7b) $$\int_{-\infty}^{\infty} H_n(x) H_m(x) \phi(x) dx = \begin{cases} n! & n = m \\ 0 & n \neq m \ . \end{cases}$$

The two useful relations

(2.1.7c) $$H_n'(x) = n H_{n-1}(x)$$

and

(2.1.7d) $$H_{n+1}(x) = x H_n(x) - n H_{n-1}(x)$$

follow immediately from the definition (2.1.7) of Hermitian polynomials. The last recursion formula may be used to determine Hermitian polynomials, starting with

$$H_0(x) = 1 \ , \quad H_1(x) = x \ .$$

An expansion of the correlation function $R_{ZZ}(t_1,t_2)$ of the output process $Z(t)$ may now be derived by inserting the expansion of the probability density function, given by expression (2.1.6), into the definition of $R_{ZZ}(t_1,t_2)$:

$$R_{ZZ}(t_1,t_2) = E\{Z(t_1)Z(t_2)\}$$
$$= \int_{-\infty}^{\infty}\int_{-\infty}^{\infty} g(u_1)g(u_2)f_{U(t_1)U(t_2)}(u_1,u_2)du_1du_2 .$$

This leads to

$$R_{ZZ}(t_1,t_2) = \sum_{n=0}^{\infty} \varrho^n(t_1,t_2)a_n(t_1)a_n(t_2) , \tag{2.1.8}$$

where

$$a_n(t_i) = \frac{1}{\sqrt{n!}\,\sigma_i}\int_{-\infty}^{\infty} g(u)H_n\left(\frac{u-m_i}{\sigma_i}\right)\phi\left(\frac{u-m_i}{\sigma_i}\right)du$$
$$= E\left\{g(U(t_i))H_n\left(\frac{U(t_i)-m_U(t_i)}{\sigma_U(t_i)}\right)\right\}/\sqrt{n!} . \tag{2.1.9}$$

These coefficients depend on m_i and σ_i and are thus for nonstationary input process $U(t)$ functions of t_i. A simple transformation of variables renders the useful representation

$$a_n(t_i) = \frac{1}{\sqrt{n!}}\int_{-\infty}^{\infty} g(v\sigma_i + m_i)H_n(v)\phi(v)dv . \tag{2.1.9a}$$

Because of $H_0(.) = 1$ there follows from Eq. (2.1.9) that

$$(2.1.10) \qquad a_0(t_i) = E\{g(U(t_i))\} = E\{Z(t_i)\}.$$

Considering the coefficients a_n as functions of m_i and σ_i and using the properties of the Hermitian polynomials, given by Eqs. (2.1.7c-d), one obtains from Eq. (2.1.9) the recursion formula

$$(2.1.9b) \qquad a_{n+1}(m_i,\sigma_i) = \frac{\sigma_i}{\sqrt{n+1}} \frac{\partial a_n(m_i,\sigma_i)}{\partial m_i}.$$

The calculation of the a_n becomes especially simple if the function g is a polynomial. Expanding g into a sum of Hermitian polynomials and using their property of orthogonality the integrations become trivial. This also shows that if g is a polynomial of nth order one has $a_k = 0$ for all $k>n$ - the sum (2.1.8) has only finitely many nonvanishing terms. In cases where infinitely many of the a_n are different from zero one has $a_n \rightarrow 0$ for $n \longrightarrow \infty$ if the mean square function of $Z(t)$ exists. This can be seen from

$$(2.1.8a) \qquad E\{Z^2(t)\} = R_{ZZ}(t,t) = \sum_{n=0}^{\infty} a_n^{\,2}(t)$$

a special case of Eq. (2.1.8).

From expansion (2.1.8) of the correlation function the analogue one for the <u>covariance function</u>

$$C_{ZZ}(t_1,t_2) = R_{ZZ}(t_1,t_2) - m_Z(t_1)m_Z(t_2)$$

$$= \sum_{n=1}^{\infty} \varrho^n(t_1,t_2)a_n(t_1)a_n(t_2) \tag{2.1.8b}$$

follows immediately because of (2.1.10).

Simple relations for the crosscorrelation and crosscovariance function are obtained by insertion of expansion (2.1.6) into the definitions of the former. Using the property of orthogonality of Hermitian polynomials one has

$$R_{ZU}(t_1,t_2) = a_0(t_1)m_2 + \varrho a_1(t_1)\sigma_2$$

$$= m_Z(t_1)m_U(t_2) + \varrho(t_1,t_2)\sigma_U(t_2)a_1(t_1) \tag{2.1.11}$$

and

$$C_{ZU}(t_1,t_2) = \varrho(t_1,t_2)\sigma_U(t_2)a_1(t_1)$$

$$= \frac{a_1(t_1)}{\sigma_U(t_1)}C_{UU}(t_1,t_2) \; . \tag{2.1.11a}$$

In cases of stationary input process $m_1 = m_2$ and $\sigma_1 = \sigma_2$ and all do not depend on time, whereas ϱ depends on $t_1 - t_2$ only. There follows that the a_n are independent of time and that

(2.1.8c) $$R_{ZZ}(t_1,t_2) = R_{ZZ}(t_1 - t_2) = \sum_{n=0}^{\infty} \varrho^n(t_1 - t_2)a_n^2 .$$

The first term of the sum is simply $E^2\{Z(t)\}$, and the second, $a_1^2\varrho(t_1-t_2)$, coincides in form with the covariance function of the input process. The remaining terms correspond to distorsions due to the nonlinearity - for a linear function $a_n = 0$ for all $n > 1$. These distorsions are frequently small, the coefficients usually decrease quite rapidly (because of the factor $1/\sqrt{n!}$), and $|\varrho^n| \leq 1$ because of $|\varrho| \leq 1$.

In such cases, of stationary input processes, an expansion of the spectral density function, $S_{ZZ}(\omega)$, may be derived by insertion of Eq. (2.1.8c) into the definition of $S_{ZZ}(\omega)$:

(2.1.8d) $$S_{ZZ}(\omega) = \sum_{n=0}^{\infty} a_n^2 S_\varrho^{(n)}(\omega) ,$$

where

(2.1.8e) $$S_\varrho^{(n)}(\omega) = \frac{1}{2\pi}\int_{-\infty}^{\infty} \varrho^n(\tau)e^{-i\omega\tau}d\tau .$$

2.2. Equivalent systems (statistical linearization)

Of what has been said in Sect. 2.1 there follows that it is, at least in principle, always possible to determine statistical properties of output processes of memoryless systems if those of the input processes are known.

However, the models under analysis are models of engineering systems. The functions which characterize these sys tems have to be determined, in principle, by measurements and are thus known numerically only. The models on the other hand should be as simple as possible to admit of simple solutions, transparent to the influence of various parameters. Moreover, approximation methods in use for treating nonlinear differential systems often require occurring nonlinear functions to belong to a specific class.

There follows that one is frequently confronted with the problem of having to approximate memoryless systems by others which are in some optimal way equivalent.

Equivalent memoryless systems:

Let

$$Z(t) = g(U(t)) \tag{2.2.12}$$

be the input-output relationship of the system under consideration. This relationship shall now be approximated by a linear combination of n given functions $G_i(.)$, i.e. by a function of the form

$$G(U(t)) = \sum_{i=1}^{n} \lambda_i(t) G_i(U(t)) \; . \tag{2.2.13}$$

The coefficients $\lambda_i(t)$ shall be such that the mean square error

$$E\{e^2(t)\} = E\{[g(U(t)) - G(U(t))]^2\} \tag{2.2.14}$$

is a minimum for every t.

In what follows the parameter t is omitted and the summation convention is used with the indices i and j ranging from 1 through n.

Necessary conditions for the optimal coefficients, $\bar{\lambda}_i$, for which the mean square error has an interior stationary value, follow immediately from $\partial E\{e^2\}/\partial\lambda_i = 0$, $i = 1,\ldots,n$:

$$\bar{\lambda}_j E\{G_i(U)G_j(U)\} = E\{g(U)G_i(U)\}, \quad i = 1,\ldots,n \tag{2.2.15}$$

The solution of this system of equations is unique if this system is nonhomogeneous and the determinant of the symmetric matrix of moments $E\{G_iG_j\}$ is different from zero.

To show that this will be the case for suitably chosen functions $G_i(U)$, assume that there is no linear combination of the G_i which vanishes identically in the domain where $f_U(u)$ is different from zero. (*) There follows then that the quadratic form $\lambda_i\lambda_j E\{G_iG_j\}$,

$$\lambda_i\lambda_j E\{G_iG_j\} = \int_{-\infty}^{\infty}[\lambda_i G_i(u)]^2 f_U(u)du,$$

(*) If $f_U(u)$ is different from zero in the interval $[-\infty,+\infty]$, with the exception of a finite number of points, this means that the $G_i(U)$ must be linearly independent.

is positive definite and thus $\det|E\{G_i G_j\}| > 0$. The function $g(u)$ shall be approximated by a linear combination of the $G_i(u)$. These functions are certainly not properly chosen if all of them are orthogonal to $g(U)$. For suitably chosen $G_i(U)$ the right-hand side of the system of equations (2.2.15) will thus not be a null vector.

That a unique solution corresponds to a minimum of the mean square error can be shown by considering the mean square error as a function of the λ_i :

$$E\{e^2\} = E\{g^2\} - 2\lambda_i E\{gG_i\} + \lambda_i\lambda_j E\{G_i G_j\} . \qquad (2.2.14a)$$

Because of $E\{e^2\} \geqslant 0$ the quadratic form on the right-hand side is nonnegative - if there is a unique stationary value it must be a <u>minimum</u>.

An equation for the stationary values of the mean square error may be obtained by multiplying each of the equations (2.2.15) by $\bar{\lambda}_i$ and summing up of the resulting equations

$$\bar{\lambda}_i\bar{\lambda}_j E\{G_i(U)G_j(U)\} = E\{\bar{G}^2(U)\}$$

$$= \bar{\lambda}_i E\{g(U)G_i(U)\} = E\{g(U)\bar{G}(U)\} .$$

Insertion of these results into Eq. (2.2.14a), valid for any λ_i, renders then

(2.2.16) $$E\{\bar{e}^2\} = E\{g^2(U)\} - E\{\bar{G}^2(U)\} .$$

Because of $E\{e^2\}\geq 0$ and thus also $E\{\bar{e}^2\}\geq 0$ this result shows that the mean square function of the output process of the equivalent system is always less than or equal to the mean square of the output of the original system.

Another interesting special result follows from Eqs. (2.2.15) if one of the G_i equals a constant, $G_k(.)=c$ say. One then obtains from the kth equation of Eqs. (2.2.15)

(2.2.15a) $$\bar{\lambda}_j E\{G_j(U)\} = E\{\bar{G}(U)\} = E\{g(U)\} ,$$

the equivalent system reproduces the mean function. In such a case not only the mean square function but also the variance function of the output of the equivalent system is less than or equal to the corresponding function of the output of the original system.

Orthonormal functions:

The relations given above become especially simple and transparent if the functions $G_i(.)$ are orthonormal on $[-\infty,\infty]$ with weight $f_U(u)$:

$$\int_{-\infty}^{\infty} G_i(u)G_j(u)f_U(u)du =$$

$$= E\{G_i(U)G_j(U)\} = \begin{cases} 1 & i = j, \\ 0 & i \neq j, \end{cases} \qquad (2.2.17)$$

(either that they are orthonormal from beginning on or orthonormal ones have been constructed from given functions by an orthonormalization procedure).

Because of this property of orthonormality one has

$$\bar{\lambda}_i = E\{g(U)G_i(U)\}, \quad i = 1,\dots,n \qquad (2.2.18)$$

and

$$E\{\bar{G}^2(U)\} = \sum_{i=1}^{n} \bar{\lambda}_i^2 . \qquad (2.2.19)$$

From Eq. (2.2.14a) there follows that

$$E\{e^2\} = E\{g^2\} + \sum_{i=1}^{n} (\lambda_i - \bar{\lambda}_i)^2 - \sum_{i=1}^{n} \bar{\lambda}_i^2 ,$$

which shows directly that the mean square error becomes a minimum for $\lambda_i = \bar{\lambda}_i$. This minimum follows as

$$E\{\bar{e}^2\} = E\{g^2\} - \sum_{i=1}^{n} \bar{\lambda}_i^2 .$$

This result, which of course could have been obtained directly from Eqs. (2.2.16) and (2.2.19), shows that the minimal mean square error does not increase if one increases the number of functions $G_i(.)$.

Gaussian input:

From Eq. (2.1.7b) there follows that orthonormal polynomials for a Gaussian input process are given by

(2.2.20) $$G_i(U) = \frac{1}{\sqrt{(i-1)!}} H_{i-1}\left(\frac{U - m_U}{\sigma_U}\right), \quad \text{(no summation!)} .$$

One then has

(2.2.18a) $$\bar{\lambda}_i = a_{i-1} ,$$

where the a_i are given by Eq. (2.1.9), and the minimal mean square error follows as

(2.2.19a) $$E\{\bar{e}^2\} = \sum_{i=n}^{\infty} a_i^{\,2} .$$

Statistical linearization I:

Single input systems:

The relations for the equivalent linear system follow immediately from those given above. Because of the importance of this statistical linearization a different notation,

(2.2.21) $$\bar{G}(U) = \nu + \nu_1 U ,$$

will be used in this case. The optimal coefficients ν and ν_1 follow immediately from Eqs. (2.2.15) as

$$\nu_1 = [E\{Ug(U)\} - E\{g(U)\}E\{U\}]/\sigma_U^2$$
$$= Cov\{U, g(U)\}/\sigma_U^2 \qquad (2.2.22a)$$

and

$$\nu = E\{g(U)\} - \nu_1 E\{U\}$$
$$= [E\{g(U)\}E\{U^2\} - E\{Ug(U)\}E\{U\}]/\sigma_U^2 \; . \qquad (2.2.22b)$$

Sometimes it is convenient to write the expression which characterizes the equivalent linear system in the form

$$\bar{G}(U) = \nu_0 + \nu_1 U^c , \qquad (2.2.23)$$

where U^c is the centered input $(E\{U^c\} = 0)$, and where

$$\nu_0 = \nu + \nu_1 m_U = E\{\bar{G}(U)\} = E\{g(U)\} \; . \qquad (2.2.23a)$$

This last equation defines ν_0, but it also shows that the equivalent linear system reproduces the mean function.

For future use, correlation function and covariance function of the output of the equivalent linear system will be calculated next. Let $Y(t)$ be the output process of this system,

$$Y(t) = \bar{G}(U(t)) = \nu(t) + \nu_1(t)U(t) \qquad (2.2.24)$$

(2.2.24) $$= \nu_0(t) + \nu_1(t)U^c(t) .$$

Because of $E\{U^c(t)\} = 0$ and $E\{U^c(t_1)U^c(t_2)\} = C_{UU}(t_1, t_2)$ the simple equations

(2.2.25a) $$R_{YY}(t_1,t_2) = \nu_0(t_1)\nu_0(t_2) + \nu_1(t_1)\nu_1(t_2)C_{UU}(t_1,t_2) ,$$

(2.2.25b) $$C_{YY}(t_1,t_2) = \nu_1(t_1)\nu_1(t_2)C_{UU}(t_1,t_2)$$

follow.

Gaussian input

In case of a Gaussian input $U(t)$ the covariance function of the output process $Z(t)$ of the original system is given by Eq. (2.1.8b). From Eq. (2.1.9) there follows that

$$a_1(t) = Cov\{U(t), g(U(t))\}/\sigma_U(t) = \nu_1(t)\sigma_U(t) ,$$

and comparison of Eqs. (2.1.8b) and (2.2.25b) then shows that in case of a Gaussian input the covariance function of the output process of the equivalent linear system coincides with the first term of series (2.1.8b), of the expansion of the covariance function of the output process of the original system. Moreover, this comparison shows also that in the inequality

$$\sigma_Y^2(t) \leq \sigma_Z^2(t) ,$$

valid for any input process, the equality sign holds, in case of a Gaussian input, only in the trivial case of a linear original system, and it shows that for a stationary Gaussian input process, where the a_n are independent of time, one has

$$C_{YY}(t_1 - t_2) < C_{ZZ}(t_1 - t_2)$$

if $g(t_1 - t_2) > 0$. The statistical linearization, discussed above, gives in case of a Gaussian input a lower value for the variance and if additionally $g > 0$ a lower one for the covariance function of the output process.

Multiple input systems:

The extension of the statistical linearization as discussed above, for a single-input system , to multiple-input systems is straightforward. Consider, for instance, a system with n inputs

$$Z = g(U_1, \dots, U_n) \qquad (2.2.26)$$

and an equivalent linear system given by

$$\bar{G}(U_1, \dots, U_n) = \nu + \sum_{i=1}^{n} \nu_i U_i \ . \qquad (2.2.27)$$

The equation the parameters ν and ν_i , $i = 1, \dots, n$, have to satisfy such that the mean square error is minimized may be obtained similarly as above . Equating the derivatives of the mean square error with respect to the $n+1$ parameters to zero one obtains

(2.2.28a) $$\nu + \nu_i E\{U_i\} = E\{g(U_1,\dots,U_n)\} ,$$

(2.2.28b) $$\nu E\{U_j\} + \nu_i E\{U_i U_j\} = E\{g(U_1,\dots,U_n)U_j\} , \quad j = 1,\dots,n .$$

Insertion of ν from the first equation into the remainders renders the system of equations

(2.2.28c) $$\nu_i \operatorname{Cov}\{U_i,U_j\} = \operatorname{Cov}\{g(U_1,\dots,U_n),U_j\} , \quad j = 1,\dots,n ,$$

for the calculation of the parameters ν_i, $i = 1,\dots,n$.

Equivalent linear systems with memory:

In the foregoing the problem of finding to a given memoryless nonlinear system an equivalent memoryless linear system which minimizes the mean square error has been considered. A question that immediately comes into one's mind is whether a linear system with memory would yield better results than a memoryless one.

To answer this question, let the nonlinear system again be given by Eq. (2.2.12),

$$Z(t) = g(U(t)) ,$$

and let the equivalent linear system with memory be characterized for $t \geqslant t_0$ by

$$Y(t) = \nu(t) + \int_{-\infty}^{\infty} \nu_1(\tau)U(t-\tau)d\tau \,, \qquad (2.2.29)$$

where $\nu(t)$ and $\nu_1(t)$, with $\nu_1(t) = 0$ for $t < t_0$, are such that the mean square error

$$E\{e^2(t)\} = E\{[Z(t) - Y(t)]^2\}$$

is a minimum for all t. This mean square error is minimal if the difference in the mean square errors for any other system and the optimal system is positive. Let

$$Y(t) + \varepsilon X(t) \,,$$

where ε is a small number and

$$X(t) = \bar{\nu}(t) + \int_{-\infty}^{\infty} \bar{\nu}_1(\tau)U(t-\tau)d\tau \,, \qquad t \geq t_0 \,,$$

with $\bar{\nu}_1(t) = 0$ for $t < t_0$, be the response of another linear system to $U(t)$. The difference

$$\Delta E\{e^2(t)\} = E\{[Y(t) + \varepsilon X(t) - Z(t)]^2\}$$

$$- E\{[Y(t) - Z(t)]^2\}$$

$$= 2\varepsilon E\{X(t)[Y(t) - Z(t)]\} + \varepsilon^2 E\{X^2(t)\}$$

is positive for any ε only if

$$E\{X(t)[Y(t) - Z(t)]\} = 0$$

Insertion of the expression for $X(t)$ shows that this equation is satisfied for arbitrary $\bar{\nu}$ and $\bar{\nu}_1$ if the relations

$$E\{Y(t)\} = E\{Z(t)\} \tag{2.2.30a}$$

and

$$\nu(t)m_U(t-\tau_1) + \int_{-\infty}^{\infty} \nu_1(\tau_2)R_{UU}(t-\tau_1, t-\tau_2)d\tau_2 = R_{UZ}(t-\tau_1, t) \tag{2.2.30b}$$

hold for all $t \geq t_0$ and all $\tau_1 \leq t - t_0$. From the first equation, which shows again that the equivalent system reproduces the mean function exactly, the equation

$$\int_{-\infty}^{\infty} \nu_1(\tau)m_U(t-\tau)d\tau = E\{Z(t)\} - \nu(t) \tag{2.2.31a}$$

valid for all $t \geq t_0$, follows. Insertion of this result and of

$$R_{UU}(t_1, t_2) = C_{UU}(t_1, t_2) + m_U(t_1)m_U(t_2)$$

into Eq. (2.2.30b) renders then

$$\int_{-\infty}^{\infty} \nu_1(\tau_2)C_{UU}(t-\tau_1, t-\tau_2)d\tau_2 = C_{UZ}(t-\tau_1, t). \tag{2.2.31b}$$

For <u>Gaussian inputs</u> it has been shown in Sect. 2.1 that

$$C_{UZ}(t-\tau,t) = C_{ZU}(t,t-\tau) = a_1(t)C_{UU}(t-\tau,t)/\sigma_U(t) .$$

There follows that in case of Gaussian inputs the equivalent linear system which minimizes the mean square error is the memoryless one discussed before.

In the case of a stationary non-Gaussian input process the problem of finding the optimal equivalent linear system reduces to that of solving the Wiener-Hopf equation

$$\int_{-\infty}^{\infty} v(\tau_2)C_{UU}(\tau_2-\tau_1)d\tau_2 = C_{ZU}(\tau) , \quad \tau_1 \geqslant t_0 \qquad (2.2.31c)$$

with $v(t)=0$ for $t < t_0$, a problem dicussed in detail in [6], [7], to name only two references.

Statistical linearization II:

The method for finding an equivalent linear system discussed above (by minimizing the mean square error) is not the only possible one. For Gaussian input processes it has been shown that the covariance functions of the output processes of the equivalent linear systems, characterized by Eqs. (2.2.21), (2.2.22), or (2.2.23), approximate those of the original systems especially well for small ϱ (and thus for large $|t_1 - t_2|$).

An equivalent linear system where the covariance functions of the output processes approximate those of the original system especially well for small $t_1 - t_2$ is obtained by requiring that mean functions and variance functions of the two systems coincide.

Let this equivalent system be characterized by

$$Y = \bar{G}(U) = \nu^* + \nu_1^* U = \nu_0^* + \nu_1^* U^c . \tag{2.2.32}$$

From the conditions

$$\begin{aligned} E\{\bar{G}(U)\} &= E\{g(U)\} , \\ Var\{\bar{G}(U)\} &= Var\{g(U)\} , \end{aligned} \tag{2.2.33}$$

there follows that the parameters ν^*, ν_1^* and ν_0^* are given by

$$\nu_0^* = E\{g(U)\} = \nu_0 , \quad \nu^* = \nu_0^* - \nu_1^* m_U \tag{2.2.32a}$$

and

$$\begin{aligned} \nu_1^{*2} &= Var\{g(U)\}/\sigma_U^2 , \\ \nu_1^* &= \pm\sqrt{Var\{g(U)\}}/\sigma_U . \end{aligned} \tag{2.2.32b}$$

The sign of ν_1^* cannot be determined from conditions (2.2.3)! It may be specified by requiring that it agrees with the sign of ν_1, or simply by inspection (*) of the function $g(.)$.

The covariance function of the output process $Y(t)$

(*) If $g(.)$ is monotonically increasing the plus sign is the correct one, if it is monotonically decreasing the minus sign.

of this equivalent linear system is given by

$$C_{YY}(t_1,t_2) = v_1^*(t_1)v_1^*(t_2)C_{UU}(t_1,t_2) . \qquad (2.2.34)$$

Stationary Gaussian input:

For stationary Gaussian input and $\varrho > 0$ one has

$$C_{ZZ}(t_1 - t_2) = \sum_{n=1}^{\infty} a_n^{\,2} \varrho^n(t_1 - t_2)$$

$$\leq \varrho(t_1 - t_2) \sum_{n=1}^{\infty} a_n^{\,2} = \varrho(t_1 - t_2)\mathrm{Var}\{g(U)\} ,$$

and thus, with Eqs. (2.2.32b) and (2.2.34), the inequality

$$C_{ZZ}(t_1 - t_2) \leq C_{YY}(t_1 - t_2) \qquad (2.2.35)$$

follows. In this case the covariance function of the output process of this equivalent linear system is an upper bound of that of the original system.

Example:

Let $U(t)$ be a stationary Gaussian stochastic process with mean m_U and covariance function $\sigma_U^{\,2} \exp\,[-x|t_1 - t_2|]$. Mean function and covariance function of the output process $Z(t)$ of the simple nonlinearity

$$g(U(t)) = U^3(t)$$

can be determined easily using the results of Sect. 2.1. The coefficients a_i follow immediately from their definition, by expanding $g(v\sigma_U+m_U)$ into a series of Hermitian polynomials and using their property of orthogonality, as

$$a_0 = m_U(m_U^2 + 3\sigma_U^2), \quad a_1 = 3\sigma_U(m_U^2 + \sigma_U^2)$$

$$a_2 = 3\sqrt{2}\, m_U\sigma_U^2, \quad a_3 = \sqrt{6}\,\sigma_U^3, \quad a_i = 0 \quad \text{for} \quad i > 3$$

The coefficients of the equivalent linear systems are then $\nu_0 = a_0$, $\nu_1 = a_1/\sigma_U$, and $\nu_1^* = \sigma_Z/\sigma_U$, where σ_Z is given (*) by $\sigma_Z^2 = a_1^2 + a_2^2 + a_3^2$. The covariance functions of the output processes of the nonlinearity and the two equivalent linear systems for the two cases $m_U = 0$ and $m_U = \sigma_U$ are plotted in Figs. 1a,b. The function $g(U) = U^3$ and the characteristics of the equivalent linear system $\nu + \nu_1 U$ are plotted for some values of m_U and σ_U in Figs. 2a,b.

(*) Check this result by calculating σ_Z^2 directly.

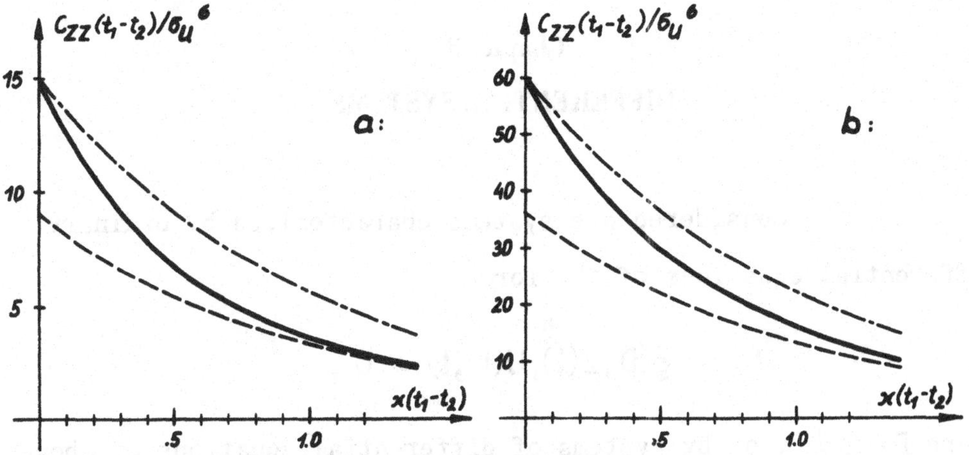

Fig. 1a, b : Covariance of $Z(t) = U^3(t)$ vs $\nu(t_1 - t_2)$.
——— exact, – – – equ. lin. system $\nu + \nu_1 U$,
–·– equ. lin. system $\nu^* + \nu_1^* U$

a: $m_u = 0$ b: $m_u = \sigma_u$

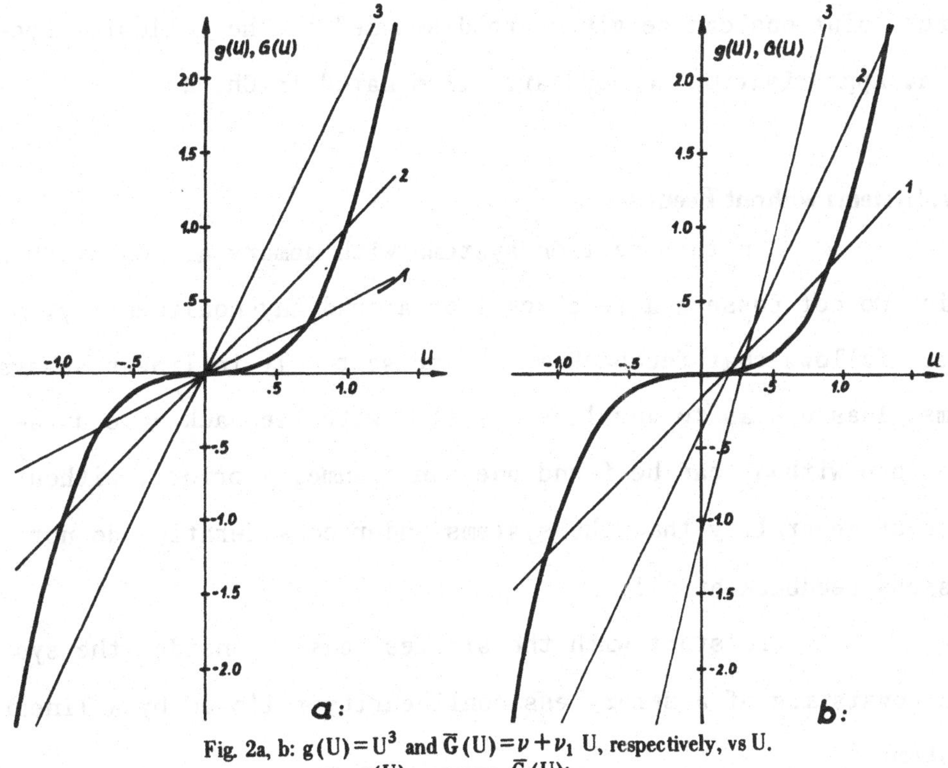

Fig. 2a, b: $g(U) = U^3$ and $\overline{G}(U) = \nu + \nu_1 U$, respectively, vs U.
——— g(U), ——— $\overline{G}(U)$;
1: $\sigma_u = 1/6$, 2: $\sigma_u^2 = 1/3$, 3: $\sigma_u^2 = 2/3$

a: $m_u = 0$ b: $m_u = \sigma_u$

Chapter 3
DIFFERENTIAL SYSTEMS

Considered are systems characterized by ordinary differential equations of the form

$$g(D, Z(t), U(t), t) = 0 ,$$

where $D = d/dt$, or by systems of differential equations of above form. Exact solultions are very rare, even if only the simplest statistical properties are required. Classes of problems where exact solutions can be given are discussed in the following section. Approximation methods are illuminated in Ch. 4.

3.1. Systems without Feedback

In this section systems with memory are considered which do not possess a feedback loop around any nonlinearity. There follows that feedback may be present only in linear subsystems. Inasmuch as to any linear system with feedback an equivalent one without can be found one can assume, a priori, without loss of generality that the systems under consideration do not possess feedback at all.

To start with the simplest case, consider the system consisting of a memoryless nonlinearity followed by a linear system:

$$Y(t) = g(U(t)) ,$$

$$Z(t) = \int_{-\infty}^{\infty} h(\tau) Y(t-\tau) d\tau , \tag{3.1.1}$$

with $h(t) = 0$ and $U(t) = 0$ for $t < t_0$. The required stochastic properties of $Y(t)$ can be determined using the results of Sect. 2.1, and the stochastic properties of the response of the whole system, $Z(t)$, then by using the theorems of the Appendix. For instance, one obtains

$$\begin{aligned} m_Y(t) &= E\{g(U(t))\} , \\ m_Z(t) &= \int_{-\infty}^{\infty} h(\tau) m_Y(t-\tau) d\tau , \end{aligned} \tag{3.1.2a}$$

or

$$\begin{aligned} R_{YY}(t_1,t_2) &= E\{g(U(t_1))g(U(t_2))\} , \\ R_{ZZ}(t_1,t_2) &= \int_{-\infty}^{\infty}\int_{-\infty}^{\infty} h(\tau_1) h(\tau_2) R_{YY}(t_1-\tau_1, t_2-\tau_2) d\tau_2 d\tau_1 , \end{aligned} \tag{3.1.2b}$$

or, if $U(t)$ and the steady-state response $Z(t)$ are stationary, one has

$$\begin{aligned} m_Z &= H(0) m_Y , \\ S_{ZZ}(\omega) &= |H(i\omega)|^2 S_{YY}(\omega) . \end{aligned} \tag{3.1.2c}$$

The situation is quite different if the two subsystems are arranged in the reverse order:

$$Y(t) = \int_{-\infty}^{\infty} h(\tau)U(t-\tau)d\tau\,,$$

$$Z(t) = g(Y(t))\,. \tag{3.1.3}$$

In order to determine first and second order moment functions of $Z(t)$ the knowledge of those of $Y(t)$ is not sufficient. In general it will be necessary to know the one-dimensional probability density function $f_{Y(t)}(y)$ in order to be able to determine mean and variance function of $Z(t)$, or the two-dimensional probability density function $f_{Y(t_1)Y(t_2)}(y_1, y_2)$ to calculate the covariance function of $Z(t)$. In cases where $g(.)$ is a polynomial (of order n) the knowledge of higher order moment functions (of order $2n$) of $Y(t)$ suffices to determine mean and covariance function of $Z(t)$. But probability density functions of the response of systems with memory are difficult to determine, and although the calculation of higher order moment functions of output processes of linear differential systems does not lead to theoretical difficulties it quite frequently leads to computational ones.

The density functions can be calculated, and the problem thus solved using the results of Sect. 2.1 and of Appendix D, if the input process $U(t)$ is Gaussian. The response $Y(t)$ of the linear subsystem is then Gaussian too and its density functions are known whenever its mean and covariance function are known.

In some cases the response of the linear subsystem to $U(t)$ can be approximated by a Gaussian process inasmuch as frequently the probability density function of the response of linear differential systems is closer to a normal one than that of the input process. Wether or not this be the case may be checked by calculating higher order moment functions of $Y(t)$ and comparing with those of a Gaussian process.

In the other cases the problem usually can be solved only by means of approximation methods, like those discussed in the following Chapter 4.

3.2. Systems with Feedback

The systems discussed in the foregoing section, differential systems without feedback, may be characterized by the fact that relationships between inputs and responses may be given in explicit form. For the systems discussed in this section, systems with memoryless nonlinearities in feedback loops or in forward branches with feedback, such a relationship in explicit form cannot be given in general.

Already in the case of the simpler systems dicussed in Sect. 3.1 it has been mentioned that frequently approximation methods have to be used whenever a nonlinear subsystem is preceeded by a system with memory the input into which is not Gaussian. But this situation will always occur in cases of nonlinear systems with feedback even if the input into the whole system is Gaussian. A nonlinear system transforms a Gaussian into a non-Gaussian process and since the input into a subsystem with feedback depends also on its output this input process is non-Gaussian.

Thus it is not surprising that practically all problems concerned with nonlinear systems with feedback do not admit of exact solutions.

There are many different methods in use which do not require an explicit relationship between input and response, and which are thus also applicable for systems with feedback. For a more or less detailed description of these methods see, for instance, [8 - 12]. The simplest ones are discussed in the following chapter.

Chapter 4
APPROXIMATION METHODS

Four simple approximation techniques for the investigation of nonlinear differential systems are discussed in the following sections. When applicable these techniques are illustrated by means of the following two examples:

Example 1:

The suspension point of the single-degree-of-freedom oscillator, shown in Fig. 3, is moved (vertically) in a stochastic manner. It is assumed that, within the range of interest, the damper is velocity-proportional, $F_d = k\dot{x}$, and that the spring characteristic may be approximated by

$$F_S = c(x + \beta x^3) . \qquad (4.0.1)$$

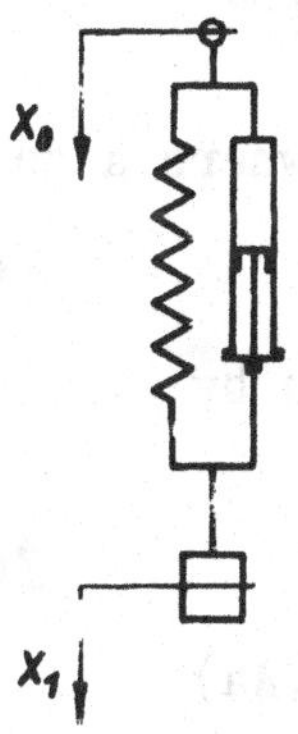

Fig. 3 : One-degree-of-freedom oscillator

The displacement of the suspension point $X_0(t)$ is measured with respect to an inertial frame and its second derivative, the acceleration of the suspension point, is assumed to be Gaussian white noise $W(t)$, with

$$E\{W(t)\} = 0 ,$$
$$E\{W(t_1)W(t_2)\} = 2D\delta(t_1 - t_2) . \qquad (4.0.2)$$

The displacement of the mass, $X_1(t)$, is measured from the equilibrium position corresponding to $X_0 = 0$. The relative displacement of the mass $Y(t) = X_1(t) - X_0(t)$ is a stochastic process the statistical properties of which we are interested in. This process may be described by the differential equation

$$\ddot{Y}(t) + 2\xi\omega_n\dot{Y}(t) + \omega_n^2(Y(t) + \beta Y^3(t)) = -W(t)\,, \tag{4.0.3}$$

where

$$2\xi\omega_n = k/m\,, \quad \omega_n^2 = c/m\,, \tag{4.0.3a}$$

and where a dot denotes differentiation with respect to time t.

With the dimensionless variables and constants given by

$$Z(\tau) = Y(t)\sqrt{\omega_n^3 D^{-1}}\,, \quad U(\tau) = W(t)\sqrt{D^{-1}\omega_n^{-1}}\,,$$
$$\tau = \omega_n t\,, \quad \lambda = \beta D\omega_n^{-3}\,. \tag{4.0.4a}$$

Eq. (4.0.3) may be rewritten in the dimensionless form

$$Z''(\tau) + 2\xi Z'(\tau) + Z(\tau) + \lambda Z^3(\tau) = -U(\tau)\,, \tag{4.0.4b}$$

where a prime denotes differentiation with respect to τ and where $U(\tau)$ is Gaussian white noise with

$$E\{U(\tau)\} = 0\,,$$
$$E\{U(\tau_1)U(\tau_2)\} = 2\delta(\tau_1 - \tau_2)\,. \tag{4.0.4c}$$

Upon usage of the notations

$$Z_1(\tau) = Z(\tau) , \qquad Z_2(\tau) = Z'(\tau) , \tag{4.0.4d}$$

the system of differential equations

$$Z_1'(\tau) = Z_2(\tau) , \qquad Z_2'(\tau) = -\left[Z_1(\tau) + \lambda Z_1^3(\tau) + 2\zeta Z_2(\tau) + U(\tau)\right] \tag{4.0.4e}$$

is equivalent to Eq.(4.0.4b).

Example 2:

The temperature of a clamped-clamped straight homogeneous beam is assumed to be uniform and given by

$$T(t) = T_r + \Theta(t) \tag{4.0.5}$$

where T_r is the reference temperature and $\Theta(t)$ denotes a stochastic fluctuation. This fluctuation is assumed to be stationary and known in the statistical sense, its mean function be identically zero.

The effect of the temperature variation is twofold: Firstly, it causes thermal stresses, and secondly, because of the temperature dependence of material parameters, variations of material properties.

The material of the beam is assumed to be viscoelastic, its constitutive law being given by Norton's law, which in the case under consideration reads

(4.0.6) $$\varepsilon = \dot{\sigma}/E + c\sigma^k + \alpha\dot{\theta}$$

where ε denotes strain, σ stress, E Young's modulus, α the modulus of linear thermal expansion, and c the coefficient of viscosity. Temperature dependence of E , α , and k usually can be neglected, but, in general, the coefficient of viscosity c shows strong temperature dependence which has to be taken into account. To account for this temperature dependence of c , but nevertheless retain a simple model it is assumed that a linear relation describes this dependence, within the range of small temperature variations we are interested in, sufficiently well:

(4.0.6a) $$c(T(t)) = c_0 + c_1\theta(t) .$$

The coefficients c_0 and c_1 are assumed to be known. For instance, they may have been calculated by statistical linearization of an experimentally determined relation $c(T)$. Of course, they will depend on the statistical properties of the temperature process then, but because of the stationarity of the latter they are at least independent of time.

For the problem under consideration $\varepsilon(t) = 0$ and the differential equation, which characterizes the behavior of the system, then reads, with Eq. (4.0.6a),

$$\ddot{\sigma}(t) + Ec_0\sigma^k(t) + Ec_1\theta(t)\sigma^k(t) = -E\alpha\dot{\theta}(t) . \quad (4.0.7a)$$

The system behavior thus may be described by a nonlinear differential equation with stochastic forcing function and a stochastic parameter. The initial state is assumed to be $\sigma(0) = 0$.

Considered are four different cases:

case a : $c_1 = 0, \; k = 1,$

case b : $c_1 \neq 0, \; k = 1,$

case c : $c_1 = 0, \; k = 3,$ and

case d : $c_1 \neq 0, \; k = 3.$

The system is linear in Case a, and nonlinear in the others.

As has been mentioned above, the statistical properties of $\theta(t)$ are assumed to be known. To be able to use the method of approximate closure of moment equations it is assumed that the process $[\theta(t), \dot{\theta}(t)]$ results by passing Gaussian white noise through the linear filter (shaping filter) given by

$$\ddot{\theta}(t) + \gamma\dot{\theta}(t) + \nu^2\theta(t) = W(t) , \quad (4.0.7b)$$

where ν^2 and γ are positive real parameters and $W(t)$ is Gaussian white noise with mean and correlation function given by Eqs.(4.0.2). The process $[\theta(t), \dot{\theta}(t)]$ then is a two-dimensional Markov process, the statistical properties of which can be determined quite easily.

Introduction of dimensionless variables and parameters given by

$$Z_1(\tau) = \sigma(t)\varkappa\ , \quad \varkappa = E^{-1}\alpha^{-1}\sqrt{\nu^3 D^{-1}}\ , \quad \tau = \nu t\ ,$$

(4.0.8)
$$Z_2(\tau) = \theta(t)\sqrt{\nu^3 D^{-1}}\ , \quad U(\tau) = W(t)\sqrt{D^{-1}\nu^{-1}}\ ,$$

$$\lambda = \gamma\nu^{-1}\ , \quad \gamma_0 = Ec_0\varkappa^{k-1}\nu^{-1}\ , \quad \gamma_1 = Ec_1\varkappa^{k-1}\sqrt{D\nu^{-5}}\ ,$$

renders Eqs. (4.0.7) in the dimensionless form

(4.0.9a) $$Z_1'(\tau) = -\gamma_0 Z_1^{\,k}(\tau) - \gamma_1 Z_1^{\,k}(\tau) Z_2(\tau) - Z_2'(\tau)\ ,$$

(4.0.9b) $$Z_2''(\tau) + \lambda Z_2'(\tau) + Z_2(\tau) = U(\tau)\ ,$$

or, with $Z_2'(\tau) = Z_3(\tau)$, in the form

(4.0.10a) $$Z_1'(\tau) = -\gamma_0 Z_1^{\,k}(\tau) - \gamma_1 Z_1^{\,k}(\tau) Z_2(\tau) - Z_3(\tau)\ ,$$
$$Z_2'(\tau) = Z_3(\tau)\ ,$$

(4.0.10b) $$Z_3'(\tau) = -Z_2(\tau) - \lambda Z_3(\tau) + U(\tau)\ .$$

As in Example 1 a prime denotes differentiation with respect to τ and $U(\tau)$ is Gaussian white noise characterized by Eqs.(4.0.4c). Equation (4.0.9a), or (4.0.10a), characterizes the system, where, as (4.0.9b) or (4.0.10b), characterizes the shaping filter (the temperature process). For simplicity of reference, the system characterized by both of Eqs. (4.0.9), or both of (4.0.10), is called the extended system.

4.1. Perturbation Technique [13 - 15]

This method, which may be used in cases where the nonlinear terms are "small" compared to the linear ones, is applied analogously as in the deterministic case.

Let, for instance, the system be characterized by the differential equation

$$N(D,t)Z(t) + \varepsilon n(D,t,U_1(t),Z(t)) = M(D,t)U_2(t) \quad (4.1.11)$$

where D stands for d/dt , $n(D,t,U_1(t),Z(t))$ is a nonlinear function, which still may depend on ε , and $N(D,t)$ and $M(D,t)$ are linear differential operators of the form

$$N(D,t) = \sum_{i=0}^{n} b_i(t)D^i ,$$
$$M(D,t) = \sum_{i=0}^{m} c_i(t)D^i , \quad (4.1.11a)$$

and where the inputs $U_1(t)$ and $U_2(t)$ may be correlated. The parameter ε is assumed to be sufficiently small such that the method is applicable.

The method rests on the assumption that $Z(t)$ permits an expansion in powers of ε ,

$$Z(t) = Z^{(0)}(t) + \varepsilon Z^{(1)}(t) + \dots , \quad (4.1.12)$$

and it is applicable in cases where the expression for n , after expansion (4.1.12) for $Z(t)$ has been inserted, may also be expanded in powers of ε,

$$
\begin{aligned}
& n(D,t,U_1(t), Z^{(0)}(t) + \varepsilon Z^{(1)}(t) + \dots) = \\
(4.1.13) \qquad & = n^{(0)}(D,t,U_1(t),Z^{(0)}(t)) \\
& + \varepsilon n^{(1)}(D,t,U_1(t),Z^{(0)}(t), Z^{(1)}(t)) + \dots ,
\end{aligned}
$$

where the $n^{(i)}$ are functions of their arguments but are independent of ε. Insertion of the two expansions (4.1.12) and (4.1.13) into Eq. (4.1.11) and comparison of equal powers of ε leads to the system of differential equations

$$
\begin{aligned}
& N(D,t)Z^{(0)}(t) = M(D,t)U_2(t) , \\
(4.1.14) \qquad & N(D,t)Z^{(1)}(t) = -n^{(0)}(D,t,U_1(t),Z^{(0)}(t)) , \\
& \vdots
\end{aligned}
$$

It can be seen that one general term in expansion (4.1.12), $Z^{(i)}(t)$ say, may be considered as linear response to an input which is a nonlinear function of t, of $U_2(t)$, possibly of derivatives of $U_2(t)$, and the previously calculated processes $Z^{(0)}(t),\dots,Z^{(i-1)}(t)$ and, possibly, of their derivatives. These differential equations can thus be solved, at least in principle, one after the other, but usually the calculations become very laborious if one goes beyond the first two equations of the sequence. For instance note that for a Gaussian input $U_2(t)$ the process $Z^{(0)}(t)$ is also Gaussian but $Z^{(1)}(t)$ is not anymore.

The statistical properties of $Z(t)$ then can be de

termined approximately from those of the $Z^{(i)}(t)$. For instance, taking expectation of expansion (4.1.12) one obtains

$$m_Z(t) = m_{Z^{(0)}}(t) + \varepsilon m_{Z^{(1)}}(t) + O(\varepsilon^2), \qquad (4.1.15a)$$

and insertion of (4.1.12) into the definition of the correlation function of $Z(t)$ renders

$$R_{ZZ}(t_1,t_2) = R_{Z^{(0)}Z^{(0)}}(t_1,t_2) + \varepsilon R_{Z^{(0)}Z^{(1)}}(t_1,t_2) + \varepsilon R_{Z^{(1)}Z^{(0)}}(t_1,t_2) + O(\varepsilon^2). \qquad (4.1.15b)$$

In order to determine the correlation function of $Z(t)$ one thus needs not only the autocorrelation functions of the $Z^{(i)}(t)$ but also their crosscorrelation functions.

Inasmuch as usually one can obtain only low-order approximations of the stochastic parameters of $Y(t)$ the practical applicability of this method is not determined by the convergence of expansions when the number of terms goes to infinity but by the asymptotic properties of these expansions for a fixed number of terms when $\varepsilon \to 0$. Special caution is appropriate in cases where the impulse response function corresponding to $N(D,t)$ is not absolutely integrable over $[0,\infty]$, see, for instance, Example 2, Cases c and d.

Example 1:

Upon assumption that λ in the differential equation (4.0.4b) of the oscillator is sufficiently small, the system of differential equations (4.1.14) reads here

$$
(4.1.16) \qquad
\begin{aligned}
&Z^{(0)''}(\tau) + 2\xi Z^{(0)'}(\tau) + Z^{(0)}(\tau) = -U(\tau)\,,\\
&Z^{(1)''}(\tau) + 2\xi Z^{(1)'}(\tau) + Z^{(1)}(\tau) = -\lambda Z^{(0)3}(\tau)\,,\\
&\qquad\vdots
\end{aligned}
$$

The impulse function

$$
(4.1.17) \qquad h(\tau)\begin{cases} = \dfrac{1}{\sqrt{1-\xi^2}}\exp(-\xi\tau)\sin\sqrt{1-\xi^2}\,\tau\,, & \tau > 0\,,\\ = 0\,, \quad \tau < 0\,, \end{cases}
$$

which corresponds to the differential operator of the left-hand sides of (4.1.16), or to the linear oscillator ($\lambda = 0$), is absolutely integrable on $[0,\infty]$ - the response of the linear oscillator to a stationary input will, after a sufficiently long time, reach stationarity. And only the steady-state behavior of the oscillator is considered.

From Eqs. (4.1.16) there follows the system of equations

$$
(4.1.16a) \qquad
\begin{aligned}
&Z^{(0)}(\tau) = -\int_0^\infty h(\eta)U(\tau-\eta)d\eta\,,\\
&Z^{(1)}(\tau) = -\lambda\int_0^\infty h(\eta)Z^{(0)3}(\tau-\eta)d\eta\,,\\
&\qquad\vdots
\end{aligned}
$$

The process $Z^{(0)}(\tau)$, being the response of a linear system to a Gaussian input, is Gaussian. Its mean function vanishes and its covariance function is given, for $\tau \geq 0$, by

$$C_{Z^{(0)}Z^{(0)}}(\tau) = \sigma_{Z^{(0)}}^2 \exp(-\zeta\tau)\left[\cos\sqrt{1-\zeta^2}\,\tau + \frac{\zeta}{\sqrt{1-\zeta^2}}\sin\sqrt{1-\zeta^2}\,\tau\right],$$

where

$$\sigma_{Z^{(0)}}^2 = 1/2\zeta . \qquad (4.1.18)$$

Since $Z^{(0)}(\tau)$ is Gaussian with vanishing mean function, there follows, from the second of the Eqs. (4.1.16a), that the mean function of $Z^{(1)}(\tau)$ vanishes also.

Equation (4.1.15b) then shows that, in order to calculate the variance of $Z(\tau)$ up to terms which are linear in ε, one needs in addition to the variance of $Z^{(0)}(\tau)$ also the covariances of $Z^{(0)}(\tau)$ and $Z^{(1)}(\tau)$. From the second of Eqs.(4.1.16a) one obtains

$$E\{Z^{(0)}(\tau)Z^{(1)}(\tau)\} = -\lambda\int_0^\infty h(\eta)E\{Z^{(0)}(\tau)Z^{(0)3}(\tau-\eta)\}d\eta . \qquad (4.1.19)$$

The expectation in the integrand may be considered as crosscorrelation function between the input, $Z^{(0)}(\tau)$, and the output of a memoryless nonlinearity, the output of which is the third power of the input. Taking into account that $Z^{(0)}(\tau)$ is Gaussian with zero mean, this expectation may be expressed in terms of the statistical properties of $Z^{(0)}(\tau)$, for instance by using Eq. (2.1.11a) and a_1 as calculated in the example at the end of Sect. 2.2:

$$E\{Z^{(0)}(\tau)Z^{(0)3}(\tau-\eta)\} = 3\sigma_{Z^{(0)}}^2 C_{Z^{(0)}Z^{(0)}}(\eta) .$$

Insertion of this result into Eq. (4.1.19) and integration renders

$$E\{Z^{(0)}(\tau)Z^{(1)}(\tau)\} = -\frac{3\lambda}{2}\sigma_{Z^{(0)}}^4 ,$$

and, with Eq. (4.1.15b), one finally has

$$(4.1.20) \qquad \sigma_Z^2 = \sigma_{Z^{(0)}}^2 - 3\lambda\sigma_{Z^{(0)}}^4 + O(\lambda^2) .$$

4.2. Statistical Linearization Methods (*) [8, 9, 16-20]

These methods are based on the ideas of approximating output processes of memoryless nonlinearities by those of equivalent linear systems, as discussed in Sect. 2.2.

Consider, for instance, a system characterized by the nonlinear differential equation

$$(4.2.21) \quad N(D,t)Z(t) + n(D,t,Z(t),U_1(t)) = M(D,t)U_2(t) ,$$

(*) Frequently, these methods are called equivalent linearization methods, but it seems to be better to reserve this designation as general notion for all techniques which use equivalent linear systems (statistical linearization, harmonic linearization, etc.)

where again the two linear operators $N(D,t)$ and $M(D,t)$ are given by (4.1.11a) and the two input-processes $U_1(t)$ and $U_2(t)$ may be correlated.

For simplicity in referencing the solution $Z^{(0)}(t)$ of the differential equation

$$N(D,t)Z^{(0)}(t) = M(D,t)U_2(t) \qquad (4.2.21a)$$

will be called the response of the corresponding linear system in what follows.

The nonlinear term $n(\ )$ in Eq. (4.2.21) may be considered as memoryless (multiple-input) nonlinear system,

$$n(D,t,Z(t),U_1(t)) = g(t,V_1(t),\ldots,V_k(t)) \ , \qquad (4.2.22)$$

where $V_1,\ldots,V_k$ stand, symbolically, for Z, U_1 , and the various derivatives of the two that may occur in the expression for $n(\_)$.

According to what has been said in Sect. 2.2 the function g , and thus n , may be approximated by the linear expression

$$\nu(t) + \nu_1(t)V_1(t) + \ldots + \nu_k(t)V_k(t) \ . \qquad (4.2.22a)$$

Insertion of this approximation into Eq. (4.2.21) renders the differential equation of the equivalent linear system:

$$N(D,t)Y(t) + \nu(t) + \nu_i(t)V_i(t) = M(D,t)U_2(t) \ . \qquad (4.2.23)$$

In this equation the summation convention is used with i ranging from 1 through k , and $Z(t)$ has been replaced by $Y(t)$ to indicate that the solution of this equation is an approximation of the solution of Eq. (4.2.21) only. Of course, this replacement will usually also be necessary for some of the V_i (which correspond to $Z(t)$ or its derivatives).

This linear system may now be used to calculate approximations for required statistical parameters of the probability law of $Z(t)$, for instance by using the results of Appendix D.

In cases where n depends on $Z(t)$, and possibly t , only the linear system equivalent to n , in the sense of statistical linearization I, may be given by

$$\nu(t) + \nu_1(t)Z(t) , \tag{4.2.22b}$$

and the differential equation which characterizes the equivalent linear system then reads

$$N(D,t)Y(t) + \nu(t) + \nu_1(t)Y(t) = M(D,t)U_2(t) . \tag{4.2.23a}$$

In this case statistical linearization method II is also applicable. One then has

$$N(D,t)Y(t) + \nu^*(t) + \nu_1^*(t)Y(t) = M(D,t)U_2(t) . \tag{4.2.23b}$$

These two simpler systems may also be used in the general case, instead of Eq. (4.2.23), to obtain quickly rough approximations.

But note that in all of these versions for finding an equivalent linear system one needs statistical properties of the "input processes" of the nonlinearity $g(t,V_1(t),\dots,V_k(t))$ in order to calculate the functions $\nu(t)$ and $\nu_i(t)$, or $\nu^*(t)$ and $\nu_1^*(t)$. But some of these processes will correspond to $Z(t)$ and its derivatives and their statistical properties are unknown.

As a first approximation one may use the corresponding quantities of the response of the corresponding linear system. The calculations are usually quite simple, particularly in cases where $U_1(t)$ and $U_2(t)$ are Gaussian, but often lead to very rough approximations only. Equivalent systems $\nu+\nu_i V_i$ and $\nu_0+\nu_i V_i^c$, where V_i^c are the centered inputs, may lead to different results and the second even to contradictions (*).

Better approximations are obtained by using the stochastic properties of the response of the equivalent linear systems, given by Eqs. (4.2.23), or (4.2.23b). But note that these systems depend on ν and ν_i, or on ν^* and ν_1^* : This procedure leads to implicit equations for the stochastic properties of $Z(t)$ if ν and ν_i, or ν^* and ν_1^*, are eliminated or to implicit ones for ν and ν_i, or ν^* and ν_1^*, if the other unknowns are elimina-

(*) For instance, consider a special system, characterized by Eq. (4.2.21) where $b_0(t)$, defined by Eq. (4.1.11a), the mean function of $U_2(t)$ and that of $Z(t)$ vanish for $t\to\infty$, but the expectation of the nonlinearity n calculated by means of the stochastic properties of the response of the corresponding linear system is different from zero. There follows that $\nu_0^*(\infty)\neq 0$, but the equivalent linear system requires that $\nu_0^*(\infty)=0$.

ted.

Frequently these implicit equations are difficult to solve. One then may use an iteration scheme, starting with ν and ν_i, or ν^* and ν_i^*, calculated by means of stochastic properties determined for the corresponding linear system, inserting them into the differential equations of the equivalent linear systems, calculating the required stochastic properties and new ν and ν_i, or ν^* and ν_i^*, and so on. One difficulty with this approach should be mentioned: Often the implicit equations, mentioned above, possess more than one real solution and then it may be difficult to select the proper one, and the iteration scheme, on the other hand, if it converges at all, need not converge towards the proper solution.

In Sect. 2.2 it has been shown for a single-input single-output nonlinearity that the optimal equivalent linear system, which minimizes the mean square error, is the memoryless one if the input process is Gaussian. It is thus not astonishing that the statistical linearization method I applied as discussed above yields reasonably good accuracy of the results if input processes of nonlinearities are close to Gaussian. This will be the case if occurring non-Gaussian processes are "normalized" sufficiently well by linear subsystems. Details about the applicability of this method may be found in [8, 9] .

Example 1:

As in Sect. 4.1 the steady-state response will be considered only.

The nonlineariry in this example is given by λZ^3 and, with the results of Sect. 2.2 and of Appendix B, one has

$$\nu_0 = \nu_0^* = \lambda E\{Z^3\},$$

$$\nu_1 = \lambda[E\{Z^4\} - E\{Z^3\}E\{Z\}]/\sigma_Z^2,$$

$$\nu_1^* = \lambda\sqrt{E\{Z^6\} - E\{Z^3\}E\{Z^3\}}/\sigma_Z.$$

Since the input process $U(t)$ is Gaussian with zero mean the response of the corresponding linear system is also Gaussian with vanishing mean, and using this

$$\nu_0 = \nu_0^* = \nu = \nu^* = 0$$

follows for this version. If the stochastic properties of the response of the equivalent linear system are used, just formally as indicated above, one can only show that these last equations are valid for one of the solutions.That this solution is the proper one can only be decided by other means. But if one starts from these results, the responses of the equivalent linear systems have then also vanishing means and are, of course, Gaussian, the parameters ν_1 and ν_1^* follow for all different versions immediately as

$$\nu_1 = \lambda E\{Z^4\}/\sigma_Z^2 = 3\lambda\sigma_Z^2,$$

$$\nu_1^* = \lambda\sqrt{E\{Z^6\}}/\sigma_Z = \lambda\sqrt{15}\,\sigma_Z^2 .$$

The variance of the response of the equivalent linear systems with still unknown ν_1 or ν_1^*, may easily be calculated: One obtains

$$\sigma_Y^2 = \frac{1}{2\xi(1+\nu_1)} = \sigma_{Z^{(0)}}^2/(1+\nu_1) , \tag{4.2.24a}$$

and

$$\sigma_Y^2 = \frac{1}{2\xi(1+\nu_1^*)} = \sigma_{Z^{(0)}}^2/(1+\nu_1^*) , \tag{4.2.24b}$$

where the solutions of the equivalent linear systems are denoted by Y and $Z^{(0)}$ is the response of the corresponding linear system, some properties of this process have already been calculated in Example 1 at the end of Sect. 4.1. Inserting now the values of ν_1 and ν_1^* as calculated by means of the statistical properties of $Z^{(0)}$ one obtains

$$\sigma_Y^2 = \sigma_{Z^{(0)}}^2/(1+3\lambda\sigma_{Z^{(0)}}^2) , \tag{4.2.24c}$$

and

$$\sigma_Y^2 = \sigma_{Z^{(0)}}^2/(1+\lambda\sqrt{15}\,\sigma_{Z^{(0)}}^2) , \tag{4.2.24d}$$

respectively. Insertion of the values of ν_1 and ν_1^* as determined by means of the properties of the response of the equivalent linear system into Eq. (4.2.24a), or Eq. (4.2.24b), renders, on the other hand, the two implicit equations

$$\sigma_Y^2 = \sigma_Z^{(0)^2}/(1 + 3\lambda\sigma_Y^2) ,$$

$$\sigma_Y^2 = \sigma_Z^{(0)^2}/(1 + \lambda\sqrt{15}\,\sigma_Y^2) .$$

These equations can be solved and give, because of $\sigma_Y^2 > 0$,

$$\sigma_Y^2 = [\sqrt{12\lambda\sigma_Z^{(0)^2} + 1} - 1]/6\lambda , \qquad (4.2.24e)$$

and

$$\sigma_Y^2 = [\sqrt{4\sqrt{15}\lambda\sigma_Z^{(0)^2} + 1} - 1]/2\lambda\sqrt{15} , \qquad (4.2.24f)$$

respectively. Note that approximation (4.2.24c) agrees with that given by (4.1.20) for small λ up to $O(\lambda)$. The approximations giv_ en by Eqs. (4.2.24c-f) are plotted in Fig. 4. The region between the two approximations given by Eqs. (4.2.24e) and (4.2.24f) is shaded (Taking weighted means of the values of ν_1 and ν_1^* the whole region is covered).

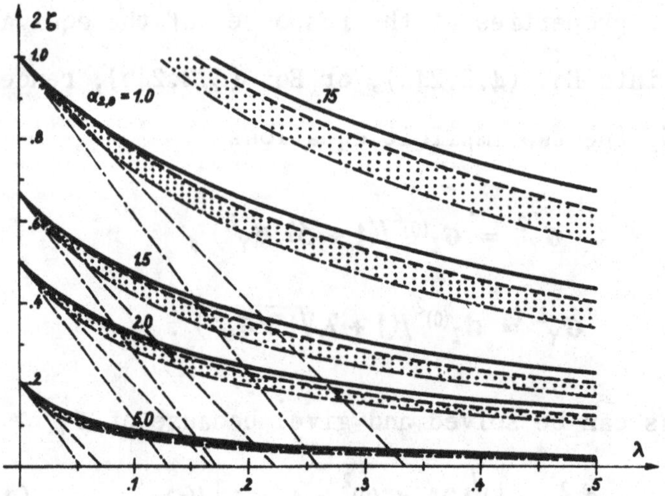

Fig. 4: Mean square relative displacement of oscillator
—— approximate closure of moment equations after moments of 4th order
– – – – Eq. (4.24e) —·— Eq. (4.24f)
– – – – Eq. (4.24c) —·— Eq. (4.24d)

4.3. Method of Approximate Closure of Moment of Equations [21]

This method bases on the theory of Markov processes, on the Markov vector approach. To be able to apply it the state of the system has to be a Markov (vector) process. In many cases the system under consideration has to be extended such that the state of this <u>extended system</u> is then Markovian, by including systems which generate the input processes, if these processes are already components of Markov vector processes, or by including the <u>shaping filters</u>, whose outputs, which are components of Markov vector processes, approximate the original system inputs, which are not. Such approximations are always possible with any desired degree of accuracy, but the construction of such shap ing filters may be a difficult problem, in particular if they have to be nonlinear.

Let a system, the state of which be already Markov ian, be characterized by the system of differential equations

$$\dot{Z}_i(t) = a_i(Z_j,t) + b_{ik}(Z_j,t)W_k(t)\ , \quad i=1,\dots,n, \qquad (4.3.25)$$

where the summation convention is used and k ranges over all, say m , input processes $W_k(t)$. These inputs are assumed to be white noise Gaussian processes with

$$\begin{aligned} E\{W_k(t)\} &= 0\ , \quad k = 1,\dots,m\ , \\ E\{W_k(t_1)W_\ell(t_2)\} &= 2D_{k\ell}\delta(t_1-t_2)\ , \quad k,\ell = 1,\dots,m\ , \end{aligned} \qquad (4.3.25a)$$

in the sense of limiting cases of ideal broad band processes. The functions a_i and b_{ik} may depend on the components $Z_j(t)$ of the state vector and possibly on time.

To the system characterized by Eqs. (4.3.25) there corresponds the differential equation

$$\frac{\partial p}{\partial t} = -\frac{\partial}{\partial z_i}(a_i p) + D_{k\ell}\frac{\partial}{\partial z_j}\left[b_{ik}\frac{\partial}{\partial z_j}(b_{j\ell}p)\right]\ , \qquad (4.3.26)$$

for the transition probability density function $p_{\underline{Z}(t)|\underline{Z}(t_0)}(\underline{z}|\underline{z}_0)$. The summation convention is used, and i and j range from 1 through n , and k and ℓ from 1 through m . This linear partial differential equation, called Fokker-Planck equation, is based on the so-called physical approach [4], which , in the case under consideration, is the appropriate one, inasmuch as the input processes $W_k(t)$ are considered as limiting cases of ideal broad band processes. Exact solutions of the Fokker-Planck equation for nonlinear problems can be given only in very rare cases, even if one restricts oneself to stationary ones. Numerical methods are applicable only if the state vector has few components and even then they are very cumbersome [22, 23]. But this differential equation may be used, under special assumptions for the functions a_i and b_{ik}, to derive differential equations for one-time moments, i.e. moments of the form $\alpha_{r_1,\dots,r_n}\{Z_1(t),\dots,Z_n(t)\}$.

Multiplication of this partial differential equation by the product $z_1^{r_1} z_2^{r_2} \dots z_n^{r_n}$, where the r_i are arbitrary nonnegative integers, and integration of the result over the whole state space renders

$$\dot{\alpha}_{r_1,\dots,r_n}\{Z_1(t),\dots,Z_n(t)\} = \sum_{i=1}^{n} r_i E\{a_i Z_i^{-1} \prod_{\nu=1}^{n} Z_\nu^{r_\nu}$$

$$+ \sum_{i,j=1}^{n}{}^{i\neq j} \sum_{k,l=1}^{m} D_{kl} r_i [r_j E\{b_{ik} b_{jl} Z_i^{-1} Z_j^{-1} \prod_{\nu=1}^{n} Z_\nu^{r_\nu}\}$$

$$+ E\{b_{ik,j} b_{jl} Z_i^{-1} \prod_{\nu=1}^{n} Z_\nu^{r_\nu}\}] \tag{4.3.27}$$

$$+ \sum_{i=1}^{n} \sum_{k,l=1}^{m} D_{kl} r_i [(r_i - 1) E\{b_{ik} b_{il} Z_i^{-2} \prod_{\nu=1}^{n} Z_\nu^{r_\nu}\}$$

$$+ E\{b_{ik,i} b_{il} Z_i^{-1} \prod_{\nu=1}^{n} Z_\nu^{r_\nu}\}],$$

where $b_{ik,j} = \partial b_{ik}/\partial Z_j$. To obtain this result integrals which involve derivatives of the transition probability density function with respect to components of the state vector have been integrated by parts and the relation

$$\dot{\alpha}_{r_1,\dots,r_n}\{Z_1(t),\dots,Z_n(t)\} = dE\{Z_1(t)^{r_1}\dots Z_n(t)^{r_n}\}/dt$$

$$= \int_{-\infty}^{\infty}\dots\int_{-\infty}^{\infty} z_1^{r_1}\dots z_n^{r_n}\,\frac{\partial p}{\partial t}\,dz_1\dots dz_n$$

has been used.

In what follows, only systems are considered where the functions a_i and b_{ik} are or have been approximated by polynomials in the components of the state vector. Under this proposition the expectations on the right-hand side of Eq. (4.3.27) may be expressed in terms of moments $\alpha_{i_1,\dots,i_n}\{Z_1(t),\dots,Z_n(t)\}$, and a linear differential equation of the form

$$\dot{\alpha}_{r_1,\dots,r_n}(t) = \sum c_{i_1,\dots,i_n}(t)\alpha_{i_1,\dots,i_n}(t) \qquad (4.3.28)$$

results. For nonlinear problems the expression on the right-hand side in general will contain also moments whose order is greater than the order of the moment on the left-hand side.

Consider systems of such differential equations obtained for all different combinations r_i such that $r_1 + \dots + r_n \leq r$. In case of a nonlinear extended system there will generally always be moments on the right-hand side of these systems whose or der is greater than r : Systems of differential equations of this type are not closed in the finite (*). There are problems where systems of differential equations (4.3.28) may be constructed which can be closed in the finite [21, 24] , but in general the systems have to be closed in the finite approximately at first in order to be able to solve them, i.e. to determine the moment functions $\alpha_{i_1,\dots,i_n}\{Z_1(t),\dots,Z_n(t)\}$ or their stationary values, if these exist.

This finite closure may be achieved by expressing moments of order greater than r approximately in terms of moments whose order is at most r (Neglection of the troublesome moments of order greater than r usually leads to meaningless results). These approximations can be obtained by means of the series expansion of the log-characteristic function in terms of semi-invariants, given by Eq. (C.8). This expansion of the log-characteristic function $\log \varphi_{Z_1(t)\dots Z_n(t)}(u_1,\dots,u_n)$ may be written in the form

$$(4.3.29) \qquad \sum_{\nu_1,\dots,\nu_n=0}^{\infty} \frac{\varkappa_{\nu_1,\dots,\nu_n}(t)}{\nu_1!\dots\nu_n!}\, i^{\Sigma\nu_k} u_1^{\nu_1}\dots u_n^{\nu_n}$$

where $i=\sqrt{-1}$ and where for simplicity of notation $\varkappa_{\nu_1,\dots,\nu_n}(t)$ stands for $\varkappa_{\nu_1,\dots,\nu_n}\{Z_1(t),\dots,Z_n(t)\}$. If one neglects now in the

(*) Conditions for the functions a_i and b_{ik} such that these systems of differential equations are or may be closed in the finite are given in [21] .

infinite sum (4.3.29) all terms which contain semi-invariants whose order is greater than r the log-characteristic function is approximated by a finite sum, denoted by $S_r(u_1,\dots,u_n,t)$:

$$S_r(u_1,\dots,u_n,t) = \tag{4.3.29a}$$

$$= \sum_{\nu_1=0}^{r} \sum_{\nu_2=0}^{r-\nu_1} \dots \sum_{\nu_n=0}^{r-\nu_1-\dots-\nu_{n-1}} \frac{\varkappa_{\nu_1,\dots,\nu_n}(t)}{\nu_1!\dots\nu_n!} i^{\Sigma\nu_k} u_1^{\nu_1}\dots u_n^{\nu_n} .$$

For the characteristic function $\varphi_{Z_1(t)\dots Z_n(t)}(u_1,\dots,u_n)$, or shortly $\varphi(u_1,\dots u_n,t)$, the approximation

$$\varphi(u_1,\dots,u_n,t) \simeq \exp[S_r(u_1,\dots,u_n,t)] \tag{4.3.30}$$

follows then.

The relationship between moment functions and characteristic function, on the other hand, is given by

$$\alpha_{j_1,\dots,j_n}(t) = i^{-\Sigma j_k} \left. \frac{\partial^{\Sigma j_k}\varphi(u_1,\dots,u_n,t)}{\partial u_1^{j_1}\dots\partial u_n^{j_n}} \right|_{u_1=\dots=u_n=0} \tag{4.3.31}$$

a generalization of Eq. (B.41). Upon insertion of approximation (4.3.30) the resulting equation may be used to approximate moment functions of order greater than r by semi-invariant functions whose order is at most r . These semi-invariant functions may then be expressed (*) in terms of moment functions, of order $\leqslant r$, by repeated usage of Eq. (4.3.31). One then obtains, via this way, approximations of moment functions of order greater than r in terms of moment functions whose order is at most r . Algorithms for $n=1$ and $n=2$ are given in [21].

This method of finite closure yields systems of differential equations, closed in the finite, but which are non-linear now. For their integration numerical methods have to be used in most of the cases. But note that these differential equations are already of a form which is especially handy for numerical integration. Problems of obtaining stationary values of moment functions lead to those of solving algebraic equations.

<u>Example 1</u>:

Inasmuch as the input $U(\tau)$ into the system described by Eqs. (4.0.4e) is a Gaussian white noise process, the process $[Z_1(\tau), Z_2(\tau)]$ is already Markovian. The Fokker-Planck equation, corresponding to Eq. (4.4.26), is given by

$$\frac{\partial p}{\partial \tau} = z_2 \frac{\partial p}{\partial z_1} + \frac{\partial}{\partial z_2}[(2\xi z_2 + z_1 + \lambda z_1^3)p] + \frac{\partial^2 p}{\partial z_2{}^2},$$

and the differential equation, corresponding to Eq.(4.3.28), reads

$$\alpha_{r,s}'(\tau) = r\alpha_{r-1,s+1}(\tau) - 2\xi s\alpha_{r,s}(\tau) - s\alpha_{r+1,s-1}(\tau)$$

(*) These expressions are exact even if approximation (4.2.30) is used.

$$-\lambda s\alpha_{r+3,s-1}(\tau) + s(s-1)\alpha_{r,s-2}(\tau)\,, \tag{4.3.32}$$

where a prime denotes differentiation with respect to τ and where $\alpha_{r,s}(\tau)$ stands for $\alpha_{r,s}\{Z_1(\tau), Z_2(\tau)\}$. If the oscillator is at rest for $\tau = 0$ the initial conditions for the moment functions are

$$\alpha_{r,s}(0) = \begin{cases} 1 & r = s = 0\,, \\ 0 & r + s \neq 0\,, \end{cases}$$

and there follows, for all positive integers k, that

$$\alpha_{2k-1,0}(\tau) = \alpha_{0,2k-1}(\tau) = 0 \quad \text{for all} \quad \tau \tag{4.3.33}$$

Consider the system of differential equations obtained from Eq. (4.3.32) for all r,s for which $r+s=2$:

$$\alpha_{2,0}'(\tau) = 2\alpha_{1,1}(\tau)\,,$$

$$\alpha_{1,1}'(\tau) = \alpha_{0,2}(\tau) - 2\zeta\alpha_{1,1}(\tau) - \alpha_{2,0}(\tau) - \lambda\alpha_{4,0}(\tau)\,, \tag{4.3.32a}$$

$$\alpha_{0,2}'(\tau) = -4\zeta\alpha_{0,2}(\tau) - 2\alpha_{1,1}(\tau) - 2\lambda\alpha_{3,1}(\tau) + 2\,.$$

The first equation simply shows that if $\alpha_{2,0}(\tau)$ reaches a stationary value, $\alpha_{2,0}'(\tau) = 0$, relative displacement and relative velocity of the oscillator are orthogonal. The two other equations contain moment functions of fourth order. If one adds to this system of differential equations the five differential equations for all r,s for which $r+s=4$ this system then contains moment functions of sixth order, and so on - these systems are not closed in the finite.

In order to determine the second order moment functions from the system (4.3.32a) the fourth order moment functions $\alpha_{4,0}(\tau)$ and $\alpha_{3,1}(\tau)$ have to be expressed approximately in terms of at most second order ones. Applying the scheme indicated above and using the results (4.3.33) one is led to

$$\alpha_{4,0}(\tau) \simeq 3\alpha_{2,0}^{2}(\tau),$$

and

$$\alpha_{3,1}(\tau) \simeq 3\alpha_{1,1}(\tau)\alpha_{2,0}(\tau).$$

Insertion of these approximations into Eqs. (4.3.32a) renders the equations

$$\alpha_{2,0}'(\tau) = 2\alpha_{1,1}(\tau),$$

$$\alpha_{1,1}'(\tau) \simeq \alpha_{0,2}(\tau) - 2\zeta\alpha_{1,1}(\tau) - \alpha_{2,0}(\tau) - 3\lambda\alpha_{2,0}^{2}(\tau),$$

$$\alpha_{0,2}'(\tau) \simeq -4\zeta\alpha_{0,2}(\tau) - 2\alpha_{1,1}(\tau) - 6\lambda\alpha_{1,1}(\tau)\alpha_{2,0}(\tau) + 2.$$

For the stationary values

$$\alpha_{1,1}(\infty) = 0, \quad \alpha_{0,2}(\infty) = 1/2\zeta,$$

$$3\lambda\alpha_{2,0}^{2}(\infty) + \alpha_{2,0}(\infty) - 1/2\zeta = 0,$$

and thus

$$\alpha_{2,0}(\infty) = [\sqrt{6\lambda/\zeta + 1} - 1]/6\lambda ,$$

follows. This result agrees with Eq. (4.3.24e), which has been obtained by means of statistical linearization. The stationary value $\alpha_{2,0}(\infty)$ has also been calculated approximately by means of finite closure of the system of differential equations for $r + s = 2$ and $r + s = 4$, after moments of fourth order, and the result is plotted in Fig. 4. All the results have been calculated for $\lambda \geqslant 0$ only inasmuch as the factor β in the characteristic of the spring has been assumed to be a fixed parameter, for a given spring, neglecting the fact that the assumed characteristic is an approximation of the actual characteristic of the spring and thus β will depend on the statistical properties of the input process. For $\lambda < 0$ there are regions in the assumed characteristic of the spring where the force exerted by the spring is not restoring anymore, and this difference is critical then.

As conclusion of this example consider the problem of determining the power required to maintain this motion. The rate at which energy is dissipated by damping is given, in terms of the dimensionless variables, by

$$D = 2\xi Z'^2$$

and the power of internal forces

$$P^{(i)} = -dV/dt - D$$

follows, where V is the potential of the spring forces. Let the power of the external forces, acting at the suspension point, and the kinetic energy be denoted by $P^{(e)}$ and T, respectively. The principle of work in its differentiated form then reads

$$\frac{dT}{dt} + \frac{dV}{dt} + D = P^{(e)} .$$

Taking expectations and interchanging the order of taking expectation and differentiation one obtains

$$\frac{d}{dt} E\{T\} + \frac{d}{dt} E\{V\} + E\{D\} = E\{P^{(e)}\} .$$

For the steady state response the expectation of V is constant, its derivative vanishes. The same is valid for the expectation of the kinetic energy if the motion of the suspension point is such that the mean square of its velocity reaches a stationary value (*). In this case one has

$$E\{P^{(e)}\} = E\{D\} = 2\zeta E\{Z'^2\} .$$

(*) f the acceleration of the suspension point were ideal Gaussian white noise the velocity is a Wiener process - a nonstationary process which never reaches stationarity.

4.4. Hierarchy Techniques [25]

These techniques are applicable only for a very restricted class of problems and they give approximations for the mean function of the response process only.

Let the system be characterized by

$$[N(D,t) + U_1(t)]Z(t) = U_2(t) , \tag{4.4.34}$$

where the two input processes $U_1(t)$ and $U_2(t)$again may be correlated and where the linear operator $N(D,t)$ is given by the first of Eqs. (4.1.11a). Multiplication of Eq. (4.4.34) by $U_1(t_1)$, $U_1(t_1)\,U_1(t_2)$, and so on, respectively, renders the system of differential equations

$$\begin{aligned} [N(D,t) + U_1(t)]Z(t) &= U_2(t) , \\ [N(D,t) + U_1(t)]U_1(t_1)Z(t) &= U_1(t_1)U_2(t) , \\ &\vdots \end{aligned} \tag{4.4.34a}$$

Following Richardson [25] this system is called the unaveraged hierarchy. Taking expectations and interchanging the order of taking expectation and differentiation one obtains the system of differential equations

$$\begin{aligned} N(D,t)E\{Z(t)\} + E\{U_1(t)Z(t)\} &= E\{U_2(t)\} , \\ N(D,t)E\{U_1(t_1)Z(t)\} + E\{U_1(t)U_1(t_1)Z(t)\} &= E\{U_1(t_1)U_2(t)\} , \end{aligned} \tag{4.4.34b}$$

called the averaged hierarchy. As in the case of the systems of differential equations for the moments, discussed in Sect. 4.3, this system is not closed in the finite. In order to close it in the finite several methods are in use: In the cumulant discard technique the system of the first r differential equations of the averaged hierarchy is closed in the finite by expressing occurring moment function of order $r+1$ approximately by those of at most rth order by neglecting all semi-invariants of order greater than r in the expansion of the log-characteristic function in terms of semi-invariants, similarly to the technique applied in Sect. 4.3. The least mean square error technique is applied on the equations of the unaveraged hierarchy and it involves the approximation of the troublesome term, which leads to the moment function of order $r+1$ in the averaged hierarchy, by a linear combination of linear functionals of $Z(t), U_1(t_1)Z(t), \dots, U_1(t)U_2(t_1)\dots U_1(t_{r-2})Z(t)$. In the correlation discard technique the finite closure of the first r equations of the averaged hierarchy is accomplished by assuming that

$$E\{U_1(t)U_1(t_1)\dots U_1(t_{r-2})Z(t)\}$$
$$\simeq E\{U_1(t)U_1(t_1)\dots U_1(t_{r-2})\}\,E\{Z(t)\},$$

i.e. that the two factors are approximately uncorrelated. This last technique is considered further in the following, details about the two others may be found in [25] .

Application of the procedure of the correlation discard technique, as indicated above, on the first equation of the hierarchy yields the linear differential equation

$$[N(D,t) + m_{U_1}(t)]m_Z(t) = m_{U_2}(t) , \qquad (4.4.35a)$$

which may be solved to obtain a first approximation of the mean function of the response $Z(t)$. An improved approximation may be obtained by application of the correlation discard approximation on the second equation of the averaged hierarchy. One obtains

$$N(D,t)E\{U_1(t_1)Z(t)\} = E\{U_1(t_1)U_2(t)\}$$
$$- E\{U_1(t)U_1(t_1)\}m_Z(t)$$

and thus, for the steady-state response,

$$E\{U_1(t_1)Z(t)\} =$$
$$= \int_{-\infty}^{t} h(t,\tau)[E\{U_1(t_1)U_2(\tau)\} - E\{U_1(t_1)U_1(\tau)\}m_Z(\tau)]d\tau ,$$

where $h(t,\tau)$ is the impulse response function corresponding to $N(D,t)$. Insertion of this result with $t_1 = t$ into the first equation of the averaged hierarchy renders the integral-differential equation,

$$N(D,t)m_Z(t) - \int_{-\infty}^{t} h(t,\tau)E\{U_1(t)U_1(\tau)\}m_Z(\tau)d\tau =$$

$$= m_{U_2}(t) - \int_{-\infty}^{t} h(t,\tau)E\{U_1(t)U_2(\tau)\}d\tau. \tag{4.4.35b}$$

Unfortunately this equation can be solved in very rare cases only. But frequently it may be used to obtain an approximation for the stationary value of $m_Z(t)$, if it exists. If for large values of t the product of the first two factors of the integrand of the integral on the left-hand side decreases, as a function of τ, much faster than $m_Z(\tau)$ does the latter may be considered, for this integration, approximately as constant and the equation

$$m_Z(\infty)\left[a_0(\infty) - \int_{-\infty}^{t} h(t,\tau)E\{U_1(t)U_1(\tau)\}d\tau\Big|_{t\to\infty}\right] =$$

$$= m_{U_2}(\infty) - \int_{-\infty}^{t} h(t,\tau)E\{U_1(t)U_2(\tau)\}d\tau\Big|_{t\to\infty} \tag{4.4.35c}$$

follows.

4.5. Example 2

The general description of this problem has already been given at the beginning of this chapter 4. The reason for including this rather lengthy example in this text is that most of the difficulties usually encountered when applying the approximation methods discussed in this chapter show up (multiple solutions of the equations with cases where the correct ones can be selected on physical grounds as well as those where comparison with other methods is necessary, parameter regions where solutions loose their meaning because of inappropriate approximations of occurring functions, impulse response functions of corresponding linear systems which are not absolutely integrable on $[0,\infty]$, and so on).

To keep this section as short as possible most details of the calculations have been omitted. Filling of these gaps may be used as an exercise.

For simplicity of notation, moment functions $\alpha_{p,r,s}\{Z_1(\tau),Z_2(\tau),Z_3(\tau)\}$ and correlation functions $R_{Z_iZ_j}(\tau_1,\tau_2)$ are denoted by $\alpha_{p,r,s}(\tau)$ and $R_{ij}(\tau_1,\tau_2)$, respectively. As usual, a prime denotes differentiation with respect to τ .

Case a:

In this simple case, discussed already in the literature [22], the system is linear, and the techniques discussed in Appendix D may be used to obtain statistical characteristics of the steady-state solution. For instance, one obtains

$$R_{22}(\tau) = A\exp(-\lambda\tau/2)\left[\cos\alpha\tau + \frac{\lambda}{\alpha}\sin\alpha\tau\right] \qquad (4.5.36a)$$

and

$$R_{11}(\tau) = B\Big[\exp(-\lambda\tau/2)\Big[(1+\gamma_0^2)\cos\alpha\tau +$$

(4.5.36b) $$+\frac{\lambda(1-\gamma_0^2)}{2\alpha}\sin\alpha\tau\Big]-\lambda\gamma_0\exp(-\gamma_0\tau)\Big],$$

where

$$A = 1/\lambda, \quad B = 1/[\lambda(1+\gamma_0(\lambda+\gamma_0))(1-\gamma_0(\lambda-\gamma_0))],$$

(4.5.36c) $$\alpha = \sqrt{4-\lambda^2}/2 .$$

All mean functions of the steady-state solution vanish.

The <u>Markov vector approach</u>, as discussed in Appendix D, also is directly applicable. The differential equation of the moment functions reads

(4.5.37) $$\alpha_{p,r,s}' = -\gamma_0 p\alpha_{p,r,s} - p\alpha_{p-1,r,s-1} + L(\alpha_{i,j,k}),$$

where the term $L(\alpha_{i,j,k})$, which corresponds to the shaping filter, is given by

(4.5.37a) $$L = r\alpha_{p,r-1,s-1} - s\alpha_{p,r+1,s-1} - \lambda s\alpha_{p,r,s} + s(s-1)\alpha_{p,r,s-2}.$$

From the system of differential equations obtained from (4.5.37) for all nonnegative integers p,r,s for which $p+r+s=2$ the stationary values

$$\alpha_{0,2,0}(\infty) = \alpha_{0,0,2}(\infty) = 1/\lambda, \quad \alpha_{0,1,1}(\infty) = 0,$$

(4.5.36d) $$\alpha_{2,0,0}(\infty) = -\alpha_{1,1,0}(\infty) = 1/[\lambda(1+\gamma_0(\lambda+\gamma_0))],$$

$$\alpha_{1,0,1}(\infty) = -\gamma_0/[\lambda(1+\gamma_0(\lambda+\gamma_0))]$$

follow immediately. Stationary values of higher order moment functions if required for the following cases, may be obtained from those above quite easily inasmuch as all processes are Gaussian. For instance, with Eq. (B. 50), the results

$$\alpha_{2,1,0} = 0\,, \quad \alpha_{2,2,0} = \alpha_{2,0,0}\alpha_{0,2,0} + 2\alpha_{1,1,0}^{\,2} \qquad (4.5.36e)$$

follow.

Case b:

Inasmuch as the Markov vector approach has been used above, the method of approximate closure of moment equations is used first. The general differential equation of the moment functions reads

$$\alpha_{p,r,s}{}' = -\gamma_0 p\alpha_{p,r,s} - \gamma_1 p\alpha_{p,r+1,s} - p\alpha_{p-1,r,s+1} + L(\alpha_{i,j,k})\,, \qquad (4.5.38)$$

where again $L(\alpha_{i,j,k})$ is given by Eq. (4.5.37a). The equations for $p=0$, which correspond to the shaping filter only, may be treated separately, the results are given by the corresponding ones of Case a. Approximate closure of the equation for $p=1$, $r=s=0$, after moment functions of first order, renders the very same result as in Case a. Finite closure of the equations for all p,r,s for which $p \neq 0$ and $p+r+s = 2$ with

$$\alpha_{2,1,0} \simeq 2\alpha_{1,0,0}\alpha_{1,1,0}\,, \quad \alpha_{1,2,0} \simeq 2\alpha_{0,1,0}\alpha_{1,1,0}\,,$$

$$\alpha_{1,1,1} \simeq \alpha_{1,0,0}\alpha_{0,1,1} + \alpha_{0,1,0}\alpha_{1,0,1} + \alpha_{0,0,1}\alpha_{1,1,0}$$

yields for $\alpha_{1,1,0}(\infty)$ and $\alpha_{1,0,1}(\infty)$ the same results as in Case a, whereas the results for $\alpha_{1,0,0}(\infty)$ and $\alpha_{2,0,0}(\infty)$ are given by

(4.5.39)
$$\alpha_{1,0,0}(\infty) = \frac{\gamma_1}{\gamma_0}\frac{1}{\lambda[1+\gamma_0(\lambda+\gamma_0)]},$$
$$\alpha_{2,0,0}(\infty) = \frac{1-2\gamma_1^2/\gamma_0^3}{\lambda[1+\gamma_0(\lambda+\gamma_0)]}.$$

The elimination process, in order to obtain the stationary values of the approximations for the moment functions, after finite closure of the equations for all p,r,s for which $p=0$ and $p+r+s=3$ is a little more complicated. But still the solution can be given in closed form. The approximations for $\alpha_{1,0,0}(\infty)$ are plotted in Fig. 5.

In order to apply the perturbation technique one has to assume that the factor γ_1 in the differential equation

(4.5.40) $$Z_1' + \gamma Z_1 + \gamma_1 Z_1 Z_2 = -Z_3,$$

see (4.0.9a), is small. Application of the technique as discussed in Sect. 4.1 renders for the steady-state response the equations

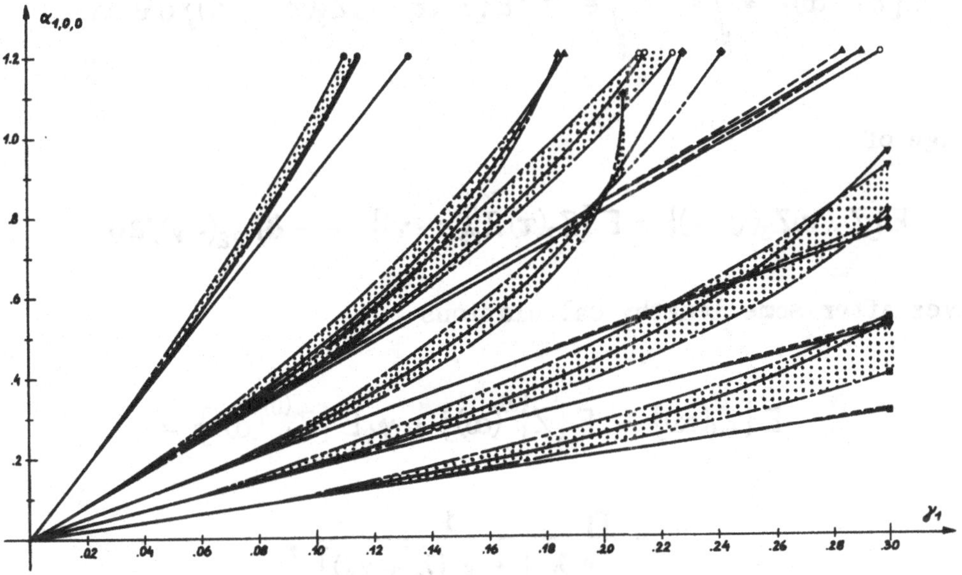

Fig. 5 : Mean value of Z_1 vs γ_1 , Case b.
——— approximate closure of moment equations (after moments of 2nd and 3rd order)
—·—·— statistical linearization ($\nu + \nu_1 Z_1$)
—··—·· statistical linearization ($\nu^* + \nu_1^* \cdot Z_1$)
– – – – hierarchy method
parameter values : ○ $\lambda = 1.0, \gamma_0 = .1$; ● $\lambda = 1.0, \gamma_0 = .2$;
▲ $\lambda = 2.0, \gamma_0 = .1$; ▼ $\lambda = 2.0, \gamma_0 = .2$; ◆ $\lambda = 3.0, \gamma_0 = .1$
■ $\lambda = 3.0, \gamma_0 = .2$

$$Z_1^{(0)} = -\int_0^\infty e^{-\gamma_0 \nu} Z_3(\tau - \nu) d\nu ,$$

$$Z_1^{(1)} = -\int_0^\infty e^{-\gamma_0 \nu} Z_2(\tau - \nu) \int_0^\infty e^{-\gamma_0 \nu_1} Z_3(\tau - \nu - \nu_1) d\nu_1 d\nu ,$$

and thus

$$E\{Z_1^{(0)}(\infty)\} = 0 ,$$

$$E\{Z_1^{(1)}(\infty)\} = \int_0^\infty e^{-\gamma_0\nu}\int_0^\infty e^{-\gamma_0\nu_1}E\{Z_2(\tau-\nu)Z_3(\tau-\nu-\nu_1)\}d\nu_1 d\nu \,.$$

Usage of

$$E\{Z_2(\tau)Z_3(\tau-\nu)\} = E\{Z_2(\tau)Z_2'(\tau-\nu)\} = -\partial R_{22}(-\nu)/\partial\nu$$

gives after some lengthy calculations

$$E\{Z_1(\infty)\} \cong E\{Z_1^{(0)}(\infty)\} + \gamma_1 E\{Z_1^{(0)}(\infty)\} =$$

$$= \frac{\gamma_1}{\gamma_0}\,\frac{1}{\lambda[1+\gamma_0(\lambda+\gamma_0)]}\,,$$

the approximation given by (4.5.39). An improvement of this result is possible but the calculations are very cumbersome.

Starting point of the statistical linearization methods is again Eq. (4.5.40), with the nonlinearity $\gamma_1 Z_1 Z_2$. If this nonlinearity is approximated by $\nu+\nu_1 Z_1+\nu_2 Z_2$ the equations for ν, ν_1 and ν_2, Eqs. (2.2.22c,d), lead to a contradiction if the statistical properties of the solution of the corresponding linear system are used to determine the required covariances, and they are complicated if those of the equivalent linear system are used. For simplicity, an equivalent system of the form $\nu+\nu_1 Z_1$ has been chosen therefore. Using the statistical properties of the response of the corresponding linear system to deter mine the required moments, see Eqs. (2.2.22a,b), one obtains the now well known approximation (4.5.39) for the stationary value of the mean function of Z_1. Usage of the statistical properties of the equivalent linear system renders (*), after some elimination, the equation

$$m_{Z_1}(\infty)[1+\gamma_0(\lambda+\gamma_0)-\gamma_1(\lambda+2\gamma_0)m_{Z_1}(\infty)+\gamma_1^2 m_{Z_1}^2(\infty)] = \tag{4.5.40}$$

$$= \gamma_1/\lambda\gamma_0 .$$

This equation possesses in some regions of the parameters λ, γ_0, and γ_1, three solutions. But only one branch of these solutions tends to zero for $\gamma_1 \to 0$, and only this branch is plotted in Fig. 5. Statistical linearization II may also be used. Using the statistical properties of the response of the corresponding linear system to calculate ν^* and ν_1^* one obtains again the approximation given by Eq. (4.5.39), whereas usage of those of the equivalent linear system renders

$$[1+(\gamma_0+m_{Z_1}(\infty))(\lambda+\gamma_0+m_{Z_1}(\infty))](\lambda m_{Z_1}^2(\infty)-1) = \tag{4.5.41}$$

$$= 1+\gamma_1^2/\lambda\gamma_0^2 .$$

(*) In calculating the required moments the property of normality of the solution and thus Eqs. (B.50) may be used. Note that these equations are valid directly for zero mean processes only. Writing $Z_1 = m_{Z_1} + Z_1^c$ they may be used directly for calculating moments which involve Z_1^c.

This equation possesses two branches of solutions with $m_{Z_1} \to 0$ as $\gamma_1 \to 0$. In Fig. 5 only that branch is plotted which is closer to the approximation (4.5.40). The region between the two approximations given by Eqs. (4.5.40) and (4.5.41) is shaded.

In this case the hierarchy technique may be applied too. Using the approximation

$$E\{Z_1(\tau)Z_2(\tau)\} \simeq E\{Z_1(\tau)\}E\{Z_2(\tau)\}$$

one obtains $m_{Z_1}(\infty) = 0$, whereas the assumption

$$E\{Z_1(\tau)Z_2(\tau)Z_2(\tau_1)\} = E\{Z_1(\tau)\}E\{Z_2(\tau)Z_2(\tau_1)\}$$

leads to the approximation

$$m_{Z_1}(\infty) = \frac{\gamma_1}{\gamma_0} \frac{1}{\lambda[1+\gamma_0(\lambda+\gamma_0)] - \gamma_1^2(\lambda+\gamma_0)} . \tag{4.5.42}$$

This result is also plotted in Fig. 5. For $\lambda = 1.0$ and $\gamma_0 = .1$ it agrees within the accuracy of drawing with approximation (4.5.39).

Case c:

The perturbation technique cannot be applied directly in this case since the impulse response function of the corresponding linear system is not absolutely integrable on $[0,\infty]$. If the technique is applied in the manner as discussed in Sect.

4.1. the process $Z_1^{(0)}(\tau)$ becomes stationary for $\tau_0 \to -\infty$, but $Z_1^{(1)}(\tau)$ does not! But rewriting the differential equation

$$Z_1'(\tau) + \gamma_0 Z_1^3(\tau) = -Z_3(\tau) \qquad (4.5.43)$$

in the form

$$Z_1' + \varepsilon Z_1 + \gamma_0 (Z_1^3 - Z_1) = -Z_3 ,$$

which agrees with Eq. (4.5.43) for $\varepsilon = \gamma_0$, the perturbation technique, in γ_0, may now be applied. Putting in the results $\varepsilon = \gamma_0$ one obtains

$$m_{Z_1}^{(0)}(\infty) = 0 ,$$

$$\sigma_{Z_1}^{(0)^2}(\infty) = 1/[\lambda(1 + \gamma_0(\lambda + \gamma_0)] ,$$

and

$$m_{Z_1}^{(1)}(\infty) = 0$$

$$\sigma_{Z_1}^{2}(\infty) = \sigma_{Z_1}^{(0)^2}(\infty)[1 - 2\gamma_0 C(3\sigma_{Z_1}^{(0)^2}(\infty) - 1)] + O(\gamma_0^2) , \qquad (4.5.44)$$

where

$$C = \frac{\lambda + \gamma_0(1 + \gamma_0^2) - \lambda[1 + \gamma_0(\lambda + \gamma_0)]/2}{1 - \gamma_0(\lambda - \gamma_0)} . \qquad (4.5.44a)$$

This approximation (4.5.44) is plotted in Fig. 6.

In applying the <u>statistical linearization method</u> I the nonlinearity $\gamma_0 Z_1^3(\tau)$ is approximated by $\nu + \nu_1 Z_1(\tau)$. From

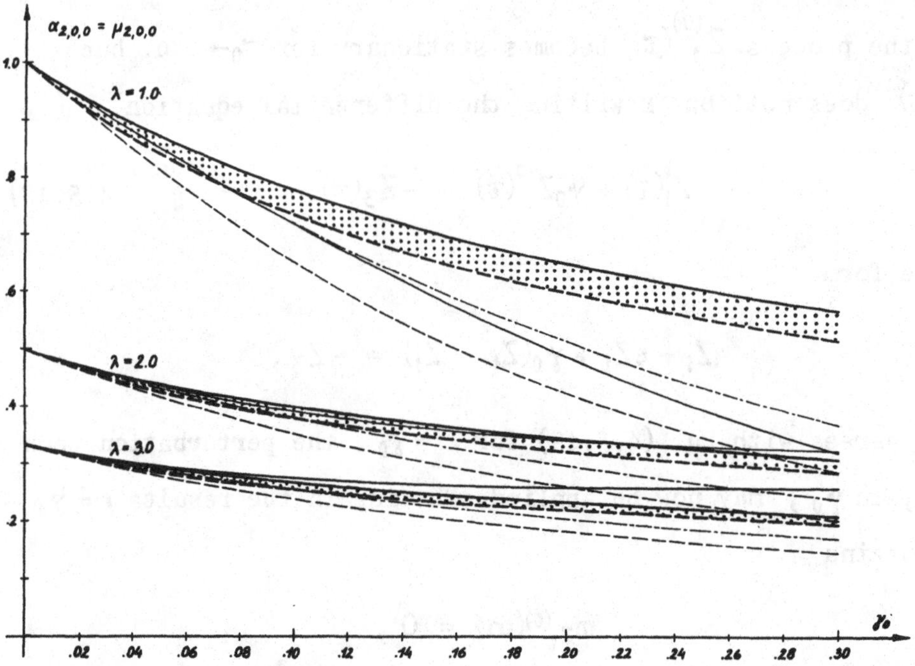

Fig. 6 : Variance of Z_1 vs γ_0, Case c.
——— Eq. (4.45b), - - - - Eq. (4.47)
——— Eq. (4.44), —·— Eq. (4.45a) - - - - Eq. (4.46)

the example in Sect. 2.2 or in Example 1.

$$\nu(\infty) = 0 , \quad \nu_1(\infty) = 3\gamma_0\sigma_{Z_1}{}^2(\infty)$$

follows. Usage of the statistical properties of the response of the corresponding linear system renders

$$m_{Z_1}(\infty) = 0 ,$$

$$\sigma_{Z_1}{}^2(\infty) = \lambda/[\lambda^2 + 3\gamma_0(\lambda^2 + 3\gamma_0)] ,$$

whereas using those of the response of the equivalent linear system one obtains the equation

$$9\gamma_0^2\sigma_{Z_1}^6(\infty)+3\gamma_0\lambda\sigma_{Z_1}^4(\infty)+\sigma_{Z_1}^2(\infty) = 1/\lambda\,. \quad (4.5.45)$$

For positive values λ and γ_0 only one real solution, plotted in Fig. 6, exists. For the parameters ν^* and ν_1^* of <u>statistical linearization method II</u>

$$\nu^*(\infty) = 0\,, \quad \nu_1^*(\infty) = 15\gamma_0^2\sigma_{Z_1}^2(\infty)$$

follows. If the statistical properties of the response of the corresponding linear system are used the approximations

$$m_{Z_1}(\infty) = 0\,, \quad \sigma_{Z_1}^2(\infty) = \lambda/[\lambda^2+\gamma_0\sqrt{15}(\lambda^2+\gamma_0\sqrt{15})]$$

follow, whereas usage of those of the response of the equivalent linear system renders

$$m_{Z_1}(\infty) = 0$$

and the equation

$$15\gamma_0^2\lambda\sigma_{Z_1}^6(\infty)+\gamma_0\lambda^2\sqrt{15}\sigma_{Z_1}^4(\infty)+\lambda\sigma_{Z_1}^2(\infty) = 1\,.$$

Starting point of the <u>method of approximation closure of moment</u> equation is the differential equation

$$\alpha_{p,r,s}{}' = -\gamma_0 p\alpha_{p+2,r,s} - p\alpha_{p-1,r,s+1} + L(\alpha_{i,j,k})\,,$$

where again $L(\alpha_{i,j,k})$ is given by (4.5.37a). It can be seen that the systems of differential equations for the moment functions can be split up into two systems, one containing only moment functions of even order, the other only those of odd order. There follows that the stationary values of all moment functions of odd order vanish. Finite closure of the equations for $p+r+s=2$, using

$$\alpha_{4,0,0} \simeq 3\alpha_{2,0,0}^2 , \quad \alpha_{3,0,1} \simeq 3\alpha_{1,0,1}\alpha_{2,0,0} ,$$

$$\alpha_{3,1,0} \simeq 3\alpha_{1,1,0}\alpha_{2,0,0} ,$$

renders again Eq. (4.5.45).

Case d:

As in Case c the perturbation technique cannot be applied directly in the manner discussed in Sect. 4.1. Considering the response $Z_1(\tau)$ of the system, which is characterized by

$$Z_1' + \gamma_0(Z_1^3 + \gamma Z_1^3 Z_2) = -Z_3 , \quad \gamma = \gamma_1/\gamma_0$$

as sum

$$Z_1 = m_{Z_1} + X ,$$

where X stands now for Z_1^c, usually used in this text, and neglecting m_{Z_1}' compared to X', and neglecting in the nonlinearity terms involving higher than first powers of m_{Z_1} compared to m_{Z_1}, one obtains

$$X' + \gamma_0 n(m_{Z_1}, X, Z_2) \simeq -Z_3 ,$$

where

$$n(m_{Z_1},X,Z_2) = (3m_{Z_1}X^2 + X^3)(1+\gamma Z_2).$$

Applying now the perturbation technique in terms of γ_0 in the same way as in Case c (adding $\varepsilon X - \gamma_0 X$) one obtains

$$X^{(0)'} + \varepsilon X^{(0)} = -Z_3 ,$$

$$X^{(1)'} + \varepsilon X^{(1)} = X^{(0)} - n^{(0)}(m_{Z_1},X^{(0)},Z_2) ,$$

where

$$n^{(0)}(m_{Z_1},X^{(0)},Z_2) = (3m_{Z_1}X^{(0)2} + X^{(0)3})(1+\gamma Z_2).$$

There follows at first that

$$\sigma_{X^{(0)}}^2(\infty) = 1/[\lambda(1+\gamma_0(\lambda+\gamma_0))] ,$$

if ε is replaced by γ_0. The expectation of $n^{(0)}(m_{Z_1}, X^{(0)}, Z_2)$ has to vanish, inasmuch as all other expectations in the equation which is obtained by taking expectation of the equation for $X^{(1)}$ vanish. This condition leads to

$$m_{Z_1}(\infty) = -\frac{\gamma E\{X^{(0)3}Z_2\}}{3E\{X^{(0)2}\}} = -\gamma E\{X^{(0)}Z_2\} ,$$

and thus to the approximation

$$m_{Z_1}(\infty) = \gamma_1/[\lambda\gamma_0(1+\gamma_0(\lambda+\gamma_0))] . \qquad (4.5.48)$$

Neglecting the term $3\gamma m_{Z_1} X^{(0)2} Z_2$ in $n^{(0)}(m_{Z_1}, X^{(0)}, Z_2)$, which is of order γ^2, one can show that only the two terms $X^{(0)}$ and $-X^{(0)3}$ contribute to the stationary value of $E\{X^{(0)}X^{(1)}\}$. There follows that the approximate result for $\sigma_X{}^2$ agrees with that of $\sigma_{Z_1}{}^2$ of Case c, Eq. (4.5.44), up to $O(\gamma_0)$. The approximation (4.5.48) for $m_{Z_1}(\infty)$ is plotted in Fig. 7.

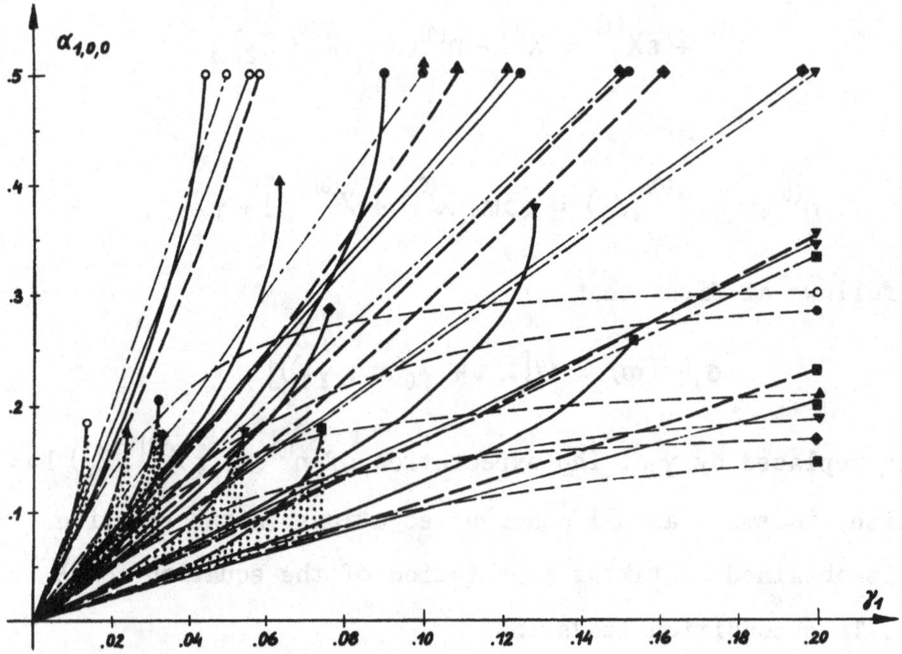

Fig. 7; Mean value of Z_1 vs γ_1, Case d.
———— Eqs. (4.51), — — — Eqs. (4.49b), – – – – Eqs. (4.50b),
———— Eq. (4.48), —·— Eqs. (4.49a), - - - - - Eqs. (4.50a).
Parameter values as in Fig. 5.

If, in applying the statistical linearization method, the nonlinearity is again approximated by $\nu + \nu_1 Z_1$ only, one obtains

$$\nu(\infty) = -3\gamma_1/\lambda^2, \quad \nu_1(\infty) = 3\gamma_0/\lambda,$$

if the statistical properties of the response of the corresponding linear system are used, and the two equations

$$\nu[1 - 2\gamma_0\nu^2/\nu_1^3 - \nu_1\nu^3/\nu_1^4 + 6\gamma_0\nu D/\nu_1^2] + 3\gamma_1 D^2 = 0 ,$$

$$\nu[9\gamma_1 D + \gamma_1\nu^2/\nu_1^2 + 3\gamma_0\nu/\nu_1]/\nu_1 + 3\gamma_0 D - \nu_1 = 0 ,$$

if those of the response of the equivalent linear system are used, where

$$D = 1/[\lambda(1 + \nu_1(\lambda + \nu_1))] ,$$

and where the argument ∞ of ν and ν_1 has been omitted. With $\nu(\infty)$ and $\nu_1(\infty)$ obtained from above equations, one has

$$m_{Z_1}(\infty) = -\nu(\infty)/\nu_1(\infty)$$

and

$$\sigma_{Z_1}^2(\infty) = D = 1/[\lambda(1 + \nu_1(\infty)(\lambda + \nu_1(\infty)))] .$$

If linearization II is applied one obtains

$$\nu^*(\infty) = -3\gamma_1/\lambda^2, \quad \nu_1^{*2}(\infty) = 15\gamma_0^2/\lambda^2 + 96\gamma_1^2/\lambda^3 ,$$

and

$$\nu^*(\nu^{*2}/\nu_1^{*2} + 3D^*)/\nu_1^* + 3\gamma_1 D^*(\nu^{*2}/\nu_1^{*2} + D^*)/\gamma_0 = 0 ,$$

$$\nu_1^{*2} = 3\gamma_0^2[3\nu^{*4}/\nu_1^{*4} + 12\nu^{*2}D^*/\nu_1^{*2} + 5D^{*2}]$$
$$+ \gamma_1^2[\nu^{*6}/\nu_1^{*6}\lambda D^* + \nu^{*4}(15/\lambda + 21D^*)/\nu_1^{*4}$$
$$+ \nu^{*2}D^*(45/\lambda + 162D^*)/\nu_1^{*2} + D^{*2}(15/\lambda + 81D^*)]$$
$$+ \gamma_0\gamma_1[6\nu^{*5}/\nu_1^{*5} + 96\nu^{*3}D^*/\nu_1^{*3} + 162\nu^* D^{*2}/\nu_1^*]$$

where $D^* = 1/[\lambda(1+\nu_1^*(\lambda + \nu_1^*))]$, if the statistical properties of the responses of the corresponding linear system and of the equivalent linear system, respectively, are used. The expressions for the approximations of $m_{Z_1}(\infty)$ and $\sigma_{Z_1}{}^2(\infty)$ follow from those given above by replacing $\nu(\infty)$ and $\nu_1(\infty)$ by $\nu^*(\infty)$ and $\nu_1^*(\infty)$, respectively. All the approximations for the stationary value of the mean of Z_1 are plotted in Fig. 7.

The general differential equation for the moment functions, used as starting point of the method of approximate closure of moment equations, is in this case given by

$$\alpha_{p,r,s}' = -\gamma_0 p\alpha_{p+2,r,s} - \gamma_1 p\alpha_{p+2,r+1,s} - p\alpha_{p-1,r,s+1} + L(\alpha_{i,j,k}) ,$$

where again $L(\alpha_{i,j,k})$ is given by (4.5.37a). Applying the technique on the equations for $p+r+s=2$ one obtains the four equations

$$3\gamma_0\alpha_{2,0,0}\alpha_{1,0,0} - 2\gamma_0\alpha_{1,0,0}{}^3 + 3\gamma_1\alpha_{2,0,0}\alpha_{1,1,0} = 0$$
$$3\gamma_0\alpha_{2,0,0}{}^2 - 2\gamma_0\alpha_{1,0,0}{}^4$$
$$+ 12\gamma_1\alpha_{1,1,0}\alpha_{1,0,0}(\alpha_{2,0,0} - \alpha_{1,0,0}{}^2) + 2\alpha_{1,0,1} = 0$$

$$(3\gamma_0\alpha_{2,0,0} + \lambda)\alpha_{1,0,1} + (6\gamma_1\alpha_{1,0,1}\alpha_{1,0,0} + 1)\alpha_{1,1,0} + 1/\lambda = 0$$

$$(3\gamma_0\alpha_{2,0,0} + 6\gamma_1\alpha_{1,1,0}\alpha_{1,0,0})\alpha_{1,1,0}$$

$$+ \gamma_1\alpha_{1,0,0}(4\alpha_{2,0,0} - 3\alpha_{1,0,0}^2)/\lambda - \alpha_{1,0,1} = 0$$

for the calculation of the stationary values $\alpha_{2,0,0}(\infty), \alpha_{1,1,0}(\infty)$, $\alpha_{1,0,1}(\infty)$ and $\alpha_{1,0,0}(\infty)$. The approximation for $\alpha_{1,0,0}(\infty)$ is again plotted in Fig. 7. For $\lambda = 3.0$, $\gamma_0 = 0.1$ and $\gamma_0 = 0.2$ the procedure used to solve the equations showed numerical instabilities for $\gamma_1 > .076$ and $\gamma_1 > 155$, respectively.

All the approximation techniques have shown the variance $\sigma_{z_1}^2(\infty) = \alpha_{2,0,0}(\infty) - \alpha_{1,0,0}^2(\infty)$ to be fairly insensitive to small variations of γ_1.

4.6. Concluding Remarks

Comparison of the four approximation methods discussed in this chapter shows that the perturbation technique and the hierarchy techniques are not very efficient. A first rough approximation may be obtained with relative ease in closed form, an improvement is very laborious. The accuracy of the second technique is usually slightly better than that of the corresponding approximation of the first but the amount of calculations is also greater.

More efficient are the other discussed methods, the statistical linearization methods and the method of approximate closure of moment equations. With the same amount of labor they show about the same accuracy. The first methods have advantages against the latter if the shaping filters are complicated, whereas the second allows for an improvement of the accuracy and thus for a numerical check of the accuracy of the approximations itself. There is another difference in these methods that should be mentioned: Inasmuch as the equivalent systems are linear the response of the original system to a Gaussian input is, in the case of the two statistical linearization methods, always approximated by Gaussian ones, the coefficients of skewness and excess of these approximations are always zero, whereas the latter method allows for these coefficients to be different from zero. Inasmuch as the method of approximate closure of moment equations leads to approximations of one-time moments only a combination with statistical linearization methods should show promise if, for instance, approximations of covariance functions of output processes are required (using the second method to determine statistical properties required for the calculation of the functions ν and ν_i, or ν^* and ν_i^*, and the first then to determine approximations of the wanted covariance functions). This approach would also avoid the difficulty of multiple solutions mentioned in Sect. 4.2.

The approximation methods have been illuminated for lumped parameter systems only. But, in order to investigate nonlinear effects in continuous parameter systems, they may be combined with the many methods in use for treating linear continuous parameter systems, like influence function methods, Ritz method and its special version the normal mode approach, the method of transfer matrices, the method of difference equations, the Holzer-Myklestad representation, and the finite element method. For details of these methods see for instance [3, 26 - 31], and the literature cited there.

LITERATURE

[1] Sweschnikow, A.A.: Untersuchungsmethoden der Theorie der Zufallsfunktionen, Leipzig: Teubner, 1965.

[2] Greensite, A.L.: Elements of Modern Control Theory, Vol. I, New York; Spartan, 1970.

[3] Parkus, H.: Random Processes in Mechanical Sciences, Udine: CISM, 1969.

[4] Gray, A.H., and T.K. Caughey: A Controversy in Problems Involving Random Parametric Excitations, J.Math. Phys. XLIV, 288 (1965).

[5] Cramér, H.: Mathematical Methods of Statistics, Princeton: Princeton University Press, 1946.

[6] Laning, J.H., and R.L. Battin: Random Processes in Automatic Control, New York: McGraw Hill, 1956.

[7] Solodovnikov, V.V.: Introduction to the Statistical Dynamics of Automatic Control Systems, New York: Dover, 1960.

[8] Pervozvanskii , A.A.: Random Processes in Nonlinear Systems, New York: Academic Press, 1965.

[9] Smith, H.W.: Approximate Analysis of Randomly Excited Nonlinear Controls, Cambridge, Mass.: M.I.T. Press, 1966.

[10] Kuznetsov, P.I., R.L. Stratonovich, and V.I. Tikhonov, Eds.: Non-Linear Transformations of Stochastic Processes, Oxford: Pergamon, 1965.

[11] Stratonovich; R.L.: Topics in the Theory of Random Noise, New York: Gordon and Breach, Vol. I, 1963, Vol. II, 1967.

[12] Deutsch, R.: Nonlinear Transformations of Random Processes, Englewood Cliffs, N.J.: Prentice-Hall, 1962.

[13] Lyon, R.H.: Response of a Nonlinear String to Random Excitation, J. Acoust. Soc. Am. 32, 953 (1960).

[14] Crandall, St. H.: Perturbation Techniques for Random Vibration of Nonlinear Systems, J. Acoust. Soc. Am. 35, 1700 (1963).

[15] Keller, J.B.: Stochastic Equations and Wave Propagation in Random Media, Proc. Symposia Appl. Math. 16, 145 (1963).

[16] Booton, R.C.: The Analysis of Nonlinear Control Systems With Random Inputs, Proc. Symposium Nonlinear Circuit Analysis, Vol. 2, New York: Polytechnic Inst. Brooklyn, 1963.

[17] Caughey, T.K.: Response of Nonlinear String to Random Loading, J. Appl. Mech., Transactions ASME 26, Ser. E, 341 (1953).

[18] Caughey, T.K.: Equivalent Linearization Techniques, J. Acoust. Soc. Am. 35, 1706 (1963).

[19] Popow, E.P., and I.P. Paltow: Nährungsmethoden zur Untersuchung nichtlinearer Regelungssysteme, Leipzig: Akademische Verlagsgesellschaft, 1963.

[20] Schlitt, H.: Theorie geregelter Systeme, Braunschweig: Viehweg, 1968.

[21] Zeman, J.L.: Zur Lösung nichtlinearer stochastischer Probleme der Mechanik, Acta Mechanica, 14, 157, (1972).

[22] Parkus, H., and J.L. Zeman: Some Stochastic Problems of Thermoviscoelasticity, In: Thermoinelasticity, Proc. IUTAM Symposium East Kilbride 1968, Edited by B.A. Boley, Wien: Springer, 1970.

[23] Merklinger, K.J.: Numerical Analysis of Non-Linear Control Systems Using the Fokker-Planck-Kolmogorov Equation, Proc. 2nd IFAC Congress, London: Butterworth, 1965.

[24] Parkus, H., and H. Bargmann: Note on the Behavior of Thermorheologically Simple Materials in Random Temperature Fields, Acta Mechanica 9, 152 (1970).

[25] Richardson, J.M.: The Application of Truncated Hierarchy Techniques in the Solution of Stochastic Differential Equations, Proc. Symposia Appl. Math. 16, (1964).

[26] Lin, Y.K.: Probabilistic Theory of Structural Dynamics, New York: McGraw-Hill, 1967.

[27] Zeman, J.L.: Oertlich und zeitlich zufällig verteilte Temperatur- und Spannungsfelder, Acta Mechanica 1,194, 371 (1965).

[28] Zeman, J.L.: A Method for the Solution of Stochastic Problems in Linear Thermoelasticity and Heat Conduction, Int. J. Solids Structures, 2, 581 (1966).

[29] Bolotin, V.V.: Statistical Methods in Structural Mechanics, San Francisco: Holden Day, 1969.

[30] Newsom, C.D., J.R. Fuller, and R.E. Sherrer: A Finite Element Approach for the Analysis of Randomly Excited, Complex, Elastic Structures, Joint AIAA-ASME Conference on Structures, Structural Dynamics and Materials, 1968, pp. 125-132.

[31] Jones, A.T., and C.W. Beadle: Random Vibration Response of Cantilever Plates Using the Finite Element Method, AIAA-Journ. 8, 1905 (1970).

APPENDIX

SOME DEFINITIONS AND THEOREMS

A. Systems:

Deterministic Systems

A <u>system</u> can be characterized by a transformation $\underline{T}[\cdot,\cdot]$ which maps the set of admissible inputs U into the set of admissible outputs Y :

$$\underline{y}(t) = \underline{T}[\underline{u}(t), \underline{s}(t_0)]\ , \ \underline{u} \in U\ , \ \underline{y} \in Y\ , \ \underline{s} \in S\ , \ t \geq t_0 > t_c , \tag{A.1}$$

where t_c is the creation time of the system, $[t_0, t]$ is the observation interval, and the <u>state</u> $\underline{s}$ of the system is the set of parameters which it is necessary to specify at t_0 such that a unique output results from any given input $\underline{u} \in U$.

Assuming that $\underline{u} \equiv \underline{0}$ is admissible, the state $\underline{\mathring{s}}$ is called a <u>zero state</u> if, for all $t_0 > t_c$, one has

$$\underline{0} = \underline{T}[\underline{0}, \underline{\mathring{s}}(t_0)]\ . \tag{A.2}$$

A state $\underline{e}$ of a system is called an <u>equilibrium state</u> if for $\underline{s}(t_0) = \underline{e}$ and $\underline{u}(t) \equiv \underline{0}$ there follows $\underline{s}(t) \equiv \underline{e}$ for all $t \geq t_0$.

The limiting terminal state g of a system when $\underline{u}(t) \equiv \underline{0}$, provided that such a limiting state exists and that it is independent of the initial state $\underline{s}(t_0)$, is called <u>ground</u>

state:

$$\underline{g} = \lim_{t\to\infty} \underline{s}(t) \quad \text{with} \quad \underline{u}(t) = 0 \quad \text{for all} \quad t \geq t_0 ,$$

$$\underline{g} \text{ independent of } \underline{s}(t_0) .$$

The response

(A.3) $$\underline{y}_0(t) = \underline{I}[\underline{u}(t), \mathring{\underline{s}}(t_0)]$$

is called zero-state response to $\underline{u}(t)$, and

$$\underline{y}_g(t) = \underline{I}[\underline{u}(t), \underline{s}(t_0) = \underline{g}]$$

is called ground-state response to $\underline{u}(t)$.

The limit of the ground-state response as $t_0 \to -\infty$, if it exists, is called the steady-state response to $\underline{u}(t)$:

(A.4) $$\underline{y}_\infty(t) = \lim_{t_0\to-\infty} \underline{I}[\underline{u}(t), \underline{s}(t_0) = \underline{g}] = \underline{I}_\infty[\underline{u}(t)] .$$

A system is said to be time-invariant if for a delayed input the output is delayed in the same way if the states at the two starting times are the same. With the delay system or delayor, defined by

$$\underline{I}_d[\underline{u}(t)] = \underline{u}(t-d) ,$$

a system is said to be time-invariant if, for all $\underline{u} \in U$, $\underline{s} \in S$, all $d > 0$, and for $t > t_c$, there is an $\underline{s}_1 \in S$ such that

(A.5) $$\underline{I}_d[\underline{I}[\underline{u}, \underline{s}(t_0)]] = \underline{I}[\underline{I}_d[\underline{u}], \underline{s}_1(t_0 + d)] ,$$

where $\underline{s}_1(t_0 + d) = \underline{s}(t_0)$.

A system is said to be <u>continuous-time</u> if $\underline{u}$ and $\underline{y}$ are defined for all real t , $t_c < t < \infty$, and it is said to be <u>discrete-time</u> if $\underline{u}$ and $\underline{y}$ are defined only for countable multiples of integers (sampling times).

A system is said to be <u>continuous-state</u> if the set of admissibles states S is a continuum, and it is said to be <u>discrete-state</u> if S is a countable set.

A system is said to be <u>memoryless</u> if, for each $t > t_c$, $\underline{y}(t)$ depends solely on $\underline{u}(t)$, and perheps t ,

$$\underline{y}(t) = \underline{g}(\underline{u}(t),t) \tag{A.6}$$

where $\underline{g}$ is a function.

A system is called <u>nonanticipative</u> if the output depends only on present and past values of its input but not on future values.

A system is said to be <u>linear</u> if

$$k\underline{T}[\underline{u}_1,\underline{s}(t_0)]-k\underline{T}[\underline{u}_2,\underline{s}(t_0)] = \underline{T}[k(\underline{u}_1-\underline{u}_2),\mathring{\underline{s}}(t_0)] \ , \tag{A.7}$$

for all $t_0 > t_c$, all $\underline{u}_1$, $\underline{u}_2 \in U$, all $\underline{s}(t_0) \in S$,and all real constants k . For $\underline{u}_2 \equiv \underline{0}$, $\underline{u}_1 = \underline{u}$, $k = 1$, the <u>fundamental decomposition theorem</u> for linear systems

$$\text{(A-8)} \qquad \underline{T}[\underline{u}, \underline{s}(t_0)] = \underline{T}[\underline{0}, \underline{s}(t_0)] + \underline{T}[\underline{u}, \underline{\mathring{s}}(t_0)]$$

follows.

Considering, for simplicity, the single-input-single-output case of a linear system, the zero-state response $T[u, \underline{\mathring{s}}(t_0)]$ is a linear functional of $u(t)$, $t \geq t_0$, then and therefore admits the formal representation (*)

$$\text{(A.9)} \qquad T[u, \underline{\mathring{s}}(t_0)] = \int_{t_0}^{t} h(t,\tau)u(\tau)d\tau .$$

Using the properties of Dirac's delta function $\delta(t)$, one can show that that the zero-state response of the system to $\delta(t-\tau)$ is given by $h(t,\tau)$ for $t \geq \tau \geq t_0$ and by zero for $\tau > t \geq t_0$. The function $h(t,\tau)$ is called <u>impulse response function</u>, therefore, and defined to be zero for $t < \tau$.

The Laplace transform of the impulse response function

$$\text{(A.10)} \qquad H(t,p) = \int_{-\infty}^{\infty} h(t,\tau)e^{-p(t-\tau)}d\tau ,$$

if it exists, is called <u>transfer function</u> or <u>frequency response function</u>. For time-invariant systems $h(t,\tau)$ and $H(t,p)$ are of the form $h(t-\tau)$ and $H(p)$, respectively.

(*) The impulse response function may contain delta functions at t_0 or t, the limits of integration are then t_0^- and t^+.

A system is called strictly proper if $h(t,\tau)$ does not contain any delta function, and it is called proper if $h(t,\tau)$ contains at most delta functions (of zero order), but does not contain delta functions of higher orders.

A system is said to be a differential system (lumped parameter system) if the relationship between inputs and outputs can be represented by a finite number of ordinary differential equations.

One of the most important concepts of system theory is that of feedback. In some analysis considerations this concept is hard to define inasmuch as for any linear system with a closed loop in its block diagram one can always find an equivalent representation without one, and vice versa. The situation is different in design of (control) systems, where one can introduce feedback purposely to decrease the sensitivity of the system to variations in parameters, to decrease the effect of nonlinearities, to stabilize the system, or to control processes automatically.

Rather intuitively one can say that a system possesses feedback (with respect to a subsystem) if a subsystem can be isolated in a forward branch of the block diagram such that the input into this subsystem depends also on its output.

Some theorems for differential systems:

An important class of single-input linear time-invariant differential systems may be described by one differential equation of the form

(A.11) $$N(D,t)y(t) = M(D,t)u(t),$$

where $N(D,t)$ and $M(D,t)$ are linear differential operators of the form

(A.12) $$N(D,t) = \sum_{i=0}^{n} a_i(t)D^i, \quad M(D,t) = \sum_{i=0}^{m} b_i(t)D^i.$$

In these equations D^i stands for d^i/dt^i and the functions a_i and b_i are assumed to be real continuous functions of time with $a_n(t) \neq 0$ for all $t \geq t_0$.

Let the impulse response fucntion of the special case $m=0$, $b_0=1$, e.g. of

(A.13) $$N(D,t)y(t) = u(t),$$

be $h_0(t,\tau)$. Per definition, this impulse response function is the solution of

$$N(D,t)y(t) = \delta(t-\tau) \text{ for all } t \text{ and } \tau \geq t_0$$

with initial conditions

$$\left.\frac{d^i y(t)}{dt^i}\right|_{t=\tau^-} = 0 \text{ for } i = 0,1,\ldots,n-1.$$

There follows that

$$N(D,t)h_0(t,\tau) = 0 \quad \text{for } t > \tau \tag{A.14}$$

and that

$$\left.\frac{\partial^i h_0(t,\tau)}{\partial t^i}\right|_{t=\tau+} = 0 \quad \text{for } i = 0,1,\ldots,n-2$$

$$\left.\frac{\partial^{n-1} h_0(t,\tau)}{\partial t^{n-1}}\right|_{t=\tau+} = \frac{1}{a_n(\tau)} \ . \tag{A.15}$$

Let $M^*\left(\frac{d}{d\tau},\tau\right)$ be the adjoint operator of M ,

$$M^*\left(\frac{d}{d\tau},\tau\right) = \sum_{i=0}^{m} (-1)^i \frac{d^i}{d\tau^i}(b_i(\tau).) \ , \tag{A.16}$$

where the dot indicates the operand. The impulse response function $h(t,\tau)$ of the general system, characterized by (A.11), is then given by

$$h(t,\tau) = M^*\left(\frac{d}{d\tau},\tau\right)h_0(t,\tau) \ . \tag{A.17}$$

The transfer function of the time-invariant system, a_i and b_i being constants then, is given by

$$H(p) = \sum_{i=0}^{m} b_i p^i / \sum_{i=0}^{n} a_i p^i \ . \tag{A.18}$$

By taking Laplace transforms of Eq. (A.11), one obtains the Laplace transform of the zero-state response y_0 to $u(t)$ and for $t_0 = 0$ as

$$\mathcal{L}\{y_0(t)\} = H(p)\mathcal{L}\{u(t)\}\,. \tag{A.19}$$

Each linear differential system may also be described by differential state equations of the form

$$\dot{\underline{s}}(t) = \underline{A}(t)\,\underline{s}(t) + \underline{B}(t)\underline{u}(t) \tag{A.20a}$$

together with an output equation of the form

$$\underline{y}(t) = \underline{C}(t)\underline{s}(t) + \underline{D}_0(t)\underline{u}(t) + \ldots + \underline{D}_k(t)\underline{u}^{(k)}(t)\,, \tag{A.20b}$$

where $\underline{u}^{(k)}$ is the kth derivative of $\underline{u}$ with respect to t. The vector $\underline{s}(t)$ is a suitable choice of the state, and, in general, the elements of the matrices $\underline{A}$, $\underline{B}$, $\underline{C}$, $\underline{D}_0$, ..., $\underline{D}_k$ are functions of time. If all of these elements are time-independent the system is time-invariant.

For convenience in reference, the two equations (A.20a) and (A.20b) together are called <u>output-state equations</u> of the differential system.

For a proper differential system the output-state equations may be reduced to

$$\begin{aligned} \dot{\underline{s}}(t) &= \underline{A}(t)\underline{s}(t) + \underline{B}(t)\underline{u}(t)\,, \\ \underline{y}(t) &= \underline{C}(t)\underline{s}(t) + \underline{D}(t)\underline{u}(t)\,. \end{aligned} \tag{A.20c}$$

In this case one has

$$\underline{s}(t) = \underline{\Phi}(t,t_0)\underline{s}(t_0) + \int_{t_0}^{t} \underline{\Phi}(t,\tau)\underline{B}(\tau)\underline{u}(\tau)d\tau , \qquad (A.21a)$$

and

$$\underline{y}(t) = \underline{C}(t)\underline{\Phi}(t,t_0)\underline{s}(t_0) + \int_{t_0}^{t} [\underline{C}(t)\underline{\Phi}(t,\tau)\underline{B}(\tau) + \underline{D}(t)\delta(t-\tau)]\underline{u}(\tau)d\tau , \qquad (A.21b)$$

where $\underline{\Phi}(t,t_0)$ is the <u>state transition matrix</u> (<u>fundamental matrix</u>). If $\underline{s}(t)$ has k components $\underline{\Phi}(t,t_0)$ is a $k \times k$ matrix which satisfies

$$\frac{d}{dt}\underline{\Phi}(t,t_0) = \underline{A}(t)\underline{\Phi}(t,t_0) , \quad \underline{\Phi}(t_0,t_0) = \underline{I}_k , \qquad (A.22)$$

$\underline{I}_k$ being the $k \times k$ identity.

If the matrices $\underline{A}$, $\underline{B}$, $\underline{C}$ and $\underline{D}$ are time-independent, the system is then time-invariant,

$$\underline{\Phi}(t,t_0) = \underline{\Phi}(t-t_0) = e^{\underline{A}(t-t_0)} , \qquad (A.23)$$

and, defining $\underline{u}(t) = \underline{0}$ for all $t < t_0$, where it is not observed, the integrals in the solutions above can be represented by convolution integrals,

$$\begin{aligned} \underline{s}(t) &= e^{\underline{A}(t-t_0)}\underline{s}(t_0) + e^{\underline{A}t}1(t) * \underline{B}\underline{u}(t) \\ \underline{y}(t) &= \underline{C}e^{\underline{A}(t-t_0)}\underline{s}(t_0) + \underline{h}(t) * \underline{u}(t) , \end{aligned} \qquad (A.24)$$

$1(t)$ being the unit-step function,

(A.25) $$\underline{h}(t) = \underline{C}e^{\underline{A}t}\underline{B}1(t) + \underline{D}\delta(t)$$

is the matrix of the impulse response functions, and the convolution integral between two functions $h(t)$ and $u(t)$ (over the interval $[-\infty,\infty]$) is defined by

(A.26) $$h(t)*u(t) = \int_{-\infty}^{\infty} h(t-\tau)u(\tau)d\tau = \int_{-\infty}^{\infty} h(\tau)u(t-\tau)d\tau .$$

The perhaps more familiar notations of convolution integrals for initial value problems, $u(t) = 0$ for all $t < t_0$, of nonanticipative systems, $h(t) = 0$ for all $t < t_0$, are

(A.27) $$h(t)*u(t) = \int_{t_0}^{t} h(t-\tau)u(\tau)d\tau = \int_{0}^{t-t_0} h(\tau)u(t-\tau)d\tau .$$

By taking Laplace transforms of the canonical state equations of the time-invariant system one obtains the Laplace transform of the zero-state response $\underline{y}_0$,

(A.28) $$\mathcal{L}\{\underline{y}_0(t)\} = \underline{H}(p)\mathcal{L}\{\underline{u}(t)\} ,$$

where

(A.29) $$\underline{H}(p) = \underline{D} + \underline{C}(p\underline{I}_k - \underline{A})^{-1}\underline{B}$$

is the transfer function matrix or frequency response function matrix, with

$$\mathcal{L}^{-1}\{H(p)\} = \underline{C}\, e^{\underline{A}t}\, \underline{B}\, 1(t) + \underline{D}\, \delta(t) = \underline{h}(t) . \qquad (A.30)$$

Random Systems

A system is said to be random if deterministic inputs $\underline{u} \in U$ result in outputs which are stochastic functions.

Literature

[1] Kalman, R.E., P.L. Falb, and M.A. Arbib: Topics in Math ematical System Theory, New York: Mc Graw-Hill, 1969.

[2] Newcomb, R.W.: Concepts of Linear Systems and Controls, Belmont, Cal.: Brooks/Cole, 1968.

[3] Zadeh, L.A., and Ch.A.Desoer: Linear System Theory, New York: McGraw-Hill, 1963.

B. Random Variables:

A (real) random variable is a real-valued quantity which has the property that for every (Borel) set B of real numbers there exists a probability $P[X \in B]$ that X is a member of B.

The probability law of a random variable X may be characterized by specifying the distribution function $F_X(.)$, defined for any real x by

$$F_X(x) = P[X \leq x]. \tag{B.1}$$

A random variable X is said to be discrete if there exists a function $p_X(.)$, the probability mass function, defined for all real numbers, such that

$$F_X(x) = \sum p_X(\xi) \tag{B.2}$$

where the sum has to be extended

over all $\xi \leq x$ with $p_X(\xi) \neq 0$.

There follows that, for any real x,

$$p_X(x) = P[X = x]. \tag{B.3}$$

A random variable X is said to be (absolutely) continuous if there exists a function $f_X(.)$, the probability density function, such that

$$F_X(x) = \int_{-\infty}^{x} f_X(\xi) d\xi. \tag{B.4}$$

There follows that

$$f_X(x) = \frac{dF_X(x)}{dx}, \tag{B.5}$$

as well as

$$P[X \in B] = \int_{\xi \in B} f_X(\xi)\,d\xi .$$

The <u>expectation of a function</u> $g(X)$ of a random variable X, $E\{g(X)\}$, is defined by

$$E\{g(X)\} = \int_{-\infty}^{\infty} g(x)\,dF_X(x) = \int_{-\infty}^{\infty} g(x) f_X(x)\,dx \quad \text{continuous case.} \tag{B.6}$$

The <u>nth (order) moment</u> of X, $\alpha_n\{X\}$, n being a positive integer, is defined by

$$\alpha_n\{X\} = E\{X^n\}, \tag{B.7}$$

and the <u>nth (order) moment about the point c</u> by $E\{(X-c)^n\}$, where, of course, c must be of the dimension of X.

The first order moment of X, m_X, is called <u>mean</u> or <u>expectation</u> of X, the second order moment <u>mean square</u> of X.

Because of

$$E\{(X-c)^2\} = E\{(X-m_X)^2\} + (c-m_X)^2 \geq E\{(X-m_X)^2\}$$

the second moment becomes a minimum when taken about the mean.

If α_n exists, the function $|X|^n$ is also integrable with respect to $F_X(.)$, so that the absolute moment

$$\beta_n\{X\} = E\{|X|^n\} \tag{B.8}$$

exists; and it then follows that α_m and β_m exist for $0 < m \leq n$.

The nth (order) moment about the mean, $\mu_n\{X\}$, is called nth (order) central moment

$$\mu_n\{X\} = E\{(X - m_X)^n\}. \tag{B.9}$$

The first central moment vanishes, and the second central moment is called variance of X:

$$\sigma_X^2 = \mathrm{Var}\{X\} = \mu_2\{X\} = E\{(X - m_X)^2\} = \alpha_2\{X\} - m_X^2, \tag{B.10}$$

where

$$\sigma_X = \sqrt{\mathrm{Var}\{X\}} \tag{B.11}$$

is called standard deviation of X.

The moment-generating function $\psi_X(.)$ of X is defined, for any real u, by

$$\psi_X(u) = E\{e^{uX}\}, \tag{B.12}$$

and the characteristic function $\varphi_X(.)$ of X, for any real u, by

$$\varphi_X(u) = E\{e^{iuX}\}. \tag{B.13}$$

A random variable may not possess a finite mean, variance, or a moment generating function, but it always possesses a characteristic function.

The relationship between the moments and the characteristic function is given by

$$\alpha_n = \frac{1}{i^n}\left(\frac{d^n \varphi_X(u)}{du^n}\right)_{u=0}, \tag{B.14}$$

or by

$$\varphi_X(u) = \sum_{n=0}^{\infty} \frac{\alpha_n}{n!}(iu)^n . \tag{B.15}$$

This last equation is valid for small arguments, for which the sum converges, but via analytical continuation one can see that $\varphi_X(.)$ is uniquely defined by the moments.

The relationship between probability mass functions, probability density functions and characteristic functions are given, for instance, by the inversion formulas

i. for any random variable X

$$P[X = x] = \lim_{U \to \infty} \frac{1}{2U} \int_{-U}^{U} e^{iux} \varphi_X(u) du , \quad -\infty < x < \infty \tag{B.16}$$

ii. for absolutely integrable $\varphi_X(.)$,

$$\int_{-\infty}^{\infty} |\varphi_X(u)| du < \infty ,$$

X is then continuous, and $f_X(.)$ is the Fourier transform of $\varphi_X(.)$,

$$f_X(x) = \frac{1}{2\pi}\int_{-\infty}^{\infty} e^{iux}\varphi_X(u)du\,, \quad -\infty < x < \infty \tag{B.17}$$

The log-characteristic function is defined as principal value of the logarithm of the characteristic function.

Defining the nth (order) cumulant or semi-invariant of X by

$$\varkappa_n\{X\} = \frac{1}{i^n}\left(\frac{d^n}{du^n}\log\varphi_X(u)\right)_{u=0}, \tag{B.18}$$

the log-characteristic function may be developed in the series for small u

$$\log\varphi_X(u) = \sum_{n=1}^{\infty}\frac{\varkappa_n}{n!}(iu)^n\,. \tag{B.19}$$

Defining the normal functions

$$\begin{aligned}\Phi(x) &= (2\pi)^{-1/2}\int_{-\infty}^{x} e^{-\xi^2/2}d\xi\,,\\ \phi(x) &= \Phi'(x) = (2\pi)^{-1/2}e^{-x^2/2}\,,\end{aligned} \tag{B.20}$$

a random variable X is said to be normal, $N[m,\sigma]$, or Gaussian, if it has the distribution function

$$F_X(x) = \Phi\left(\frac{x-m}{\sigma}\right), \tag{B.21a}$$

or equivalently if it has the density function

$$f_X(x) = \frac{1}{\sigma}\phi\left(\frac{x-m}{\sigma}\right), \qquad \text{(B.21b)}$$

or equivalently if it has the characteristic function

$$\exp\left[imu - \frac{\sigma^2 u^2}{2}\right]. \qquad \text{(B.21c)}$$

Then m is the mean and σ the standard deviation of X, and the central moments

$$\mu_n = \begin{cases} 0 & \text{for} \quad n \quad \text{odd} \\ 1.3\ldots(2k-1)\sigma^{2k} & \text{for} \quad n = 2k \end{cases} \qquad \text{(B.22)}$$

follow.

The importance of moments in applications follows from the fact that they are simple parameters which describe the main features of the distribution. To locate distributions, by finding some "typical value" which may be conceived as a "central point" of the distribution the mean may be taken. As a parameter which gives an idea of how widely the values of the variables are spread the standard deviation, the square root of μ_2, may be used. For a symmetric distribution every central moment of odd order is equal to zero. Thus the <u>coefficient of skewness</u>

$$\gamma_1 = \mu_3/\sigma^3 \qquad (B.23)$$

may be regarded as a measure of the asymmetry of the distribution (roughly spoken, γ_1 is positive for distributions with a long tail on the positive side of the mean). And as a measure of the degree of flattening of the distribution near the mean the <u>coefficient of excess</u>

$$\gamma_2 = \frac{\mu_4}{\sigma^4} - 3 \qquad (B.24)$$

may be taken.

Moreover, the importance of moments for characterizing a random variable X follows from Tchebychev's inequality

$$P[|X - m_X| \geq h\sigma_X] \leq 1/h^2, \quad h \text{ arbitrary} \qquad (B.25)$$

and from <u>Markov's inequality</u>

$$P[|X| \geq k] \leq \beta_n/k^n, \qquad (B.26)$$

for any k and any n such that β_n is finite.

The (<u>joint</u>) <u>probability distribution function</u> of two random variables X_1 and X_2 is defined for any real x_1 and x_2 by

(B.27) $$F_{X_1X_2}(x_1,x_2) = P[X_1 \leqslant x_1 \, , \, X_2 \leqslant x_2] \, .$$

There follows that

(B.28) $$F_{X_1}(x_1) = F_{X_1X_2}(x_1, \infty) \, ,$$
$$F_{X_2}(x_2) = F_{X_1X_2}(\infty, x_2) \, .$$

The (joint) probability density function of two (absolutely) continuous random variables X_1 and X_2 is defined for any real x_1 and x_2 by

(B.29) $$f_{X_1X_2}(x_1,x_2) = \frac{\partial^2 F_{X_1X_2}(x_1,x_2)}{\partial x_1 \partial x_2} \, .$$

There follows that

(B.30) $$f_{X_1}(x_1) = \int_{-\infty}^{\infty} f_{X_1X_2}(x_1,x_2)dx_2 \, ,$$
$$f_{X_2}(x_2) = \int_{-\infty}^{\infty} f_{X_1X_2}(x_1,x_2)dx_1 \, .$$

The expectation of a function $g(X_1,X_2)$ of two random variables X_1 and X_2 is defined by

(B.31) $$E\{g(X_1,X_2)\} = \int_{-\infty}^{\infty} g(x_1,x_2)d^2F_{X_1X_2}(x_1,x_2) =$$
$$= \int_{-\infty}^{\infty}\int_{-\infty}^{\infty} g(x_1,x_2)f_{X_1X_2}(x_1,x_2)dx_1dx_2 \qquad \text{continuous case} \, .$$

The __nth__ (__order__) __moments__ of two random variables X_1 and X_2, $\alpha_{r_1,r_2}\{X_1,X_2\}$ with $r_1+r_2=n$, are defined by

$$\alpha_{r_1,r_2}\{X_1,X_2\} = E\{X_1^{r_1}X_2^{r_2}\}, \quad r_1+r_2 = n, \tag{B.32}$$

and the __nth__ (__order__) __central moments__ $\mu_{r_1,r_2}\{X_1,X_2\}$, with $r_1+r_2=n$, by

$$\mu_{r_1,r_2}\{X_1,X_2\} = E\{(X_1-E\{X_1\})^{r_1}(X_2-E\{X_2\})^{r_2}\}, \quad r_1+r_2=n. \tag{B.33}$$

The two first order moments are the means of X_1 and X_2, respectively, the two first order central moments vanish, the two second order moments $\alpha_{2,0}$ and $\alpha_{0,2}$ and the two second order central moments $\mu_{2,0}$ and $\mu_{0,2}$ are the mean squares and variances of X_1 and X_2, respectively.

The (third) second order central moment $\mu_{1,1}$ is called __covariance__ of X_1 and X_2:

$$\begin{aligned} Cov\{X_1,X_2\} &= \mu_{1,1}\{X_1,X_2\} = \\ &= E\{(X_1-E\{X_1\})(X_2-E\{X_2\})\} = \\ &= E\{X_1X_2\} - E\{X_1\}E\{X_2\}. \end{aligned} \tag{B.34}$$

The analogue between second order moments of two random variables and second order moments of mass geometry follows from the definitions.

Two random variables X_1 and X_2 are said to be __uncorrelated__ if their covariance vanishes; or equivalently if the __correlation coefficient__, defined by

(B.35) $$\varrho\{X_1,X_2\} = \frac{\mathrm{Cov}\{X_1,X_2\}}{\sigma_{X_1}\sigma_{X_2}}$$

vanishes; or equivalently, assuming that X_1 and X_2 may be added, if

$$\mathrm{Var}\{X_1+X_2\} = \mathrm{Var}\{X_1\} + \mathrm{Var}\{X_2\} .$$

Two random variables X_1 and X_2 are said to be <u>orthogonal</u> if

(B.36) $$E\{X_1X_2\} = 0 .$$

A very useful theorem, <u>Schwarz's</u> or <u>Cauchy's</u> inequality, states, that for any two random variables, X_1 and X_2, with finite second moments,

(B.37) $$E^2\{X_1X_2\} = |E\{X_1X_2\}|^2 \leq E\{X_1^2\}E\{X_2^2\} ,$$

where the equality holds if and only if, for some constant c, $X_2 = cX_1$. Using condition (B.37) for the centered random variables $X_1 - E\{X_1\}$ and $X_2 - E\{X_2\}$, there follows that

(B.38) $$|\mathrm{Cov}\{X_1,X_2\}|^2 \leq \mathrm{Var}\{X_1\}\mathrm{Var}\{X_2\} .$$

The <u>moment-generating function</u> $\Psi_{X_1X_2}(.,.)$ of two random variables X_1 and X_2 is defined, for any real u_1 and u_2, by

$$\Psi_{X_1X_2}(u_1,u_2) = E\{\exp(u_1X_1 + u_2X_2)\}, \tag{B.39}$$

and the <u>characteristic</u> <u>function</u> $\varphi_{X_1X_2}(.,.)$ of X_1 and X_2 by

$$\varphi_{X_1X_2}(u_1,u_2) = E\{\exp[i(u_1X_1 + u_2X_2)]\}. \tag{B.40}$$

The relationship between the moments of two random variables and their characteristic function is given by

$$\alpha_{r_1,r_2} = i^{-r_1-r_2}\left(\frac{\partial^{r_1+r_2}}{\partial u_1^{r_1}\partial u_2^{r_2}}\varphi_{X_1X_2}(u_1,u_2)\right)_{u_1=u_2=0}, \tag{B.41}$$

and the <u>nth</u> (<u>order</u>) <u>cumulants</u>, $\varkappa_{r_1,r_2}\{X_1,X_2\}$ with $r_1 + r_2 = n$, are defined by

$$\varkappa_{r_1,r_2}\{X_1,X_2\} = i^{-r_1-r_2}\left(\frac{\partial^{r_1+r_2}}{\partial u_1^{r_1}\partial u_2^{r_2}}\log\varphi_{X_1X_2}(u_1,u_2)\right)_{u_1=u_2=0}. \tag{B.42}$$

Let $Q(u_1,u_2)$,

$$Q(u_1,u_2) = \mu_{2,0}u_1^2 + 2\mu_{1,1}u_1u_2 + \mu_{0,2}u_2^2,$$

be a positive definite quadratic form. Then the inverse of $Q(u_1,u_2)$, $Q^{-1}(x_1,x_2)$, exists and is of the form

$$Q^{-1}(x_1,x_2) = \frac{1}{1-\varrho^2}\left(\frac{x_1^2}{\sigma_1^2} - \frac{2\varrho x_1x_2}{\sigma_1\sigma_2} + \frac{x_2^2}{\sigma_2^2}\right).$$

Two continuous random variables X_1 and X_2 are

said to be jointly normally distributed if they possess the probability density function

(B.43a) $$f_{X_1X_2}(x_1,x_2) = \frac{1}{2\pi\sigma_1\sigma_2\sqrt{1-\varrho^2}}\exp\left[-\frac{1}{2}Q^{-1}(x_1-m_1,x_2-m_2)\right],$$

or, equivalently, the characteristic function

(B.43b) $$\varphi_{X_1X_2}(u_1,u_2) = \exp\left[i(m_1u_1+m_2u_2)-\frac{1}{2}Q(u_1,u_2)\right].$$

Then m_1 and m_2 are the means, σ_1 and σ_2 the standard deviations of X_1 and X_2, respectively, and ϱ their correlation coefficient.

The extension of the definitions, given above for two random variables, with the exception of the definitions of the covariance, the correlation coefficient, and of uncorrelated random variables, to any finite number of random variables is straightforward.

Jointly distributed random variables $X_1, X_2, \ldots, X_n$ are said to be independent if for all $B_1, B_2, \ldots, B_n$

(B.44a) $$P[X_1\in B_1,\ldots,X_n\in B_n] = P[X_1\in B_1]\ldots P[X_n\in B_n],$$

or equivalently if, for all real $x_1, x_2, \ldots, x_n$,

(B.44b) $$F_{X_1\ldots X_n}(x_1,\ldots,x_n) = F_{X_1}(x_1)\ldots F_{X_n}(x_n),$$

or equivalently if, for all real $u_1, u_2, \ldots, u_n$,

$$\varphi_{X_1 \ldots X_n}(u_1, \ldots, u_n) = \varphi_{X_1}(u_1) \ldots \varphi_{X_n}(u_n), \tag{B.44c}$$

or equivalently if, for all functions $g_1(.)$, $g_2(.)$, ..., $g_n(.)$ for which all the following expectations exist,

$$E\{g_1(X_1) \ldots g_n(X_n)\} = E\{g_1(X_1)\} \ldots E\{g_n(X_n)\}. \tag{B.44d}$$

There follows that two independent random variables are also uncorrelated (but not all uncorrelated ones are independent!)

Regarding n jointly distributed random variables as components of an n-dimensional random vector $\underline{X}$, with expectation $\underline{m}_X$, the n random variables are said to be jointly normally distributed if their probability density function is

$$f_{\underline{X}}(\underline{x}) = \frac{1}{(2\pi)^{n/2} |C|^{1/2}} \exp\left[-\frac{1}{2}(\underline{x} - \underline{m}_X)^T \underline{C}^{-1} (\underline{x} - \underline{m}_X)\right], \tag{B.45a}$$

or equivalently if their characteristic function is

$$\varphi_{\underline{X}}(\underline{u}) = \exp\left(i\underline{m}_X{}^T \underline{u} - \frac{1}{2}\underline{u}^T \underline{C} \underline{u}\right), \tag{B.45b}$$

where $\underline{C}$ is the matrix of covariances and variances,

$$\underline{C} = E\{(\underline{X} - \underline{m}_X)(\underline{X} - \underline{m}_X)^T\}, \tag{B.46}$$

$|C|$ the determinant of $\underline{C}$, and T indicates transposition.

The relationship between higher order moments of jointly normally distributed random variables and their moments of at most second order can be deduced quite easily from the definitions given above. For instance, inserting the characteristic function of n random variables, given by (B.45b), into

$$E\{X_1 \dots X_n\} = i^{-n}\left(\frac{\partial^n \varphi_{\underline{X}}(\underline{u})}{\partial u_1 \dots \partial u_n}\right),$$

a special case of the generalization of Eq. (B.41) to n random variables, one obtains for n normally distributed random variables with zero means

$$E\{X_1 X_2 \dots X_{2m+1}\} = 0$$

(B.47)

$$E\{X_1 X_2 \dots X_{2m}\} = \sum E\{X_i X_j\} E\{X_k X_\ell\},$$

where the sum is to be taken over all $(2m)!/(m!\,2^m)$ different ways by which we can group $2m$ elements into m pairs, and where random variables with different indices may be identical. For instance, for $X_1 = X_2 =, \dots, = X_{2m} = X$ equation (B.22) follows. In the general case of random variables with means different form zero Eqs. (B.47) are valid for central moments.

A sequence of random variables $X_1, X_2, \dots, X_n,$ is said to converge to X

i. in probability, $X_n \xrightarrow{i.pr.} X$, if for every $\varepsilon > 0$

$$P[|X_n - X| > \varepsilon] \to 0 ;$$

ii. with probability one or almost everywhere, $X_n \xrightarrow{a.e.} X$, if

$$P[\lim_{n\to\infty} X_n = X] = 1 ;$$

iii. in mean square, $X_n \xrightarrow{m.s.} X$ or $\text{l.i.m.}\, X_n = X$ ("limit in the mean") if

$$E\{(X_n - X)^2\} \to 0;$$

iv. in distribution if

$$\lim_{n\to\infty} F_{X_n}(x) = F_X(x) .$$

Convergence with probability 1 and convergence in mean square imply convergence in probability, but they do not necessarily imply each other.

The calculus based on mean square convergence, which is used throughout this booklet, is called mean square calculus.

Functions of random variables:

Let $Y_1, \ldots, Y_m$ be m random variables related to $n \geq m$ continuous random variables $X_1, \ldots, X_m$ by

$$Y_i = g_i(X_1, \ldots, X_n) , \quad i = 1, \ldots, m ,$$

where the g_i's are functions in each of the arguments.

Then the moments of the Y_i are given by

(B.48)
$$\alpha_{r_1,\ldots,r_m}\{Y_1,\ldots,Y_m\} = \int_{-\infty}^{\infty}\cdots\int_{-\infty}^{\infty} g_1^{r_1}(x_1,\ldots,x_n)\ldots g_m^{r_m}(x_1,\ldots,x_n) f_{X_1\ldots X_n}(x_1,\ldots,x_n)dx_n\ldots dx_1 \,.$$

For $m=1$, $n=2$ Eq. (B.31) follows.

The distribution function of the Y_i, $f_{Y_1\ldots Y_m}(y_1,\ldots,y_m)$, may be given by

$$\int\cdots\int f_{X_1\ldots X_n}(x_1,\ldots,x_n)dx_n\ldots dx_1$$

where the integral has to be extended over all regions for which $g_i(x_1,\ldots,x_n) \leq y_i$, $i = 1,\ldots,m$.

For $m=n$ and one-to-one functions g_i the probability density function of the Y_i is given by

(B.49)
$$f_{Y_1\ldots Y}(y_1,\ldots,y_n) = f_{X_1\ldots X_n}(x_1,\ldots,x_n)/|J(x_1,\ldots,x_n)| \,,$$

where

(B.50a)
$$y_i = g_i(x_1,\ldots,x_n), \quad i = 1,\ldots,n \,,$$

and where the determinant $J(x_1,\ldots,x_n)$ is defined by

(B.50b)
$$J(x_1,\ldots,x_n) = \begin{vmatrix} \frac{\partial g_1}{\partial x_1},\ldots,\frac{\partial g_1}{\partial x_n} \\ \cdots\cdots\cdots\cdots \\ \frac{\partial g_n}{\partial x_1},\ldots,\frac{\partial g_n}{\partial x_n} \end{vmatrix} \,.$$

This relation may also be used in the case $m<n$, introducing $n-m$ auxiliary variables $Y_{m+1},\dots,Y_n$ defined by

$$Y_i = X_i\,, \quad i = m+1,\dots,n\,.$$

Literature:

see the literature to Appendix: C. Stochastic Processes.

C. Stochastic Processes

A (scalar-valued) *stochastic process*, *random process*, or *random function* is a parametered family of random variables with the parameter, such as time, varying in an index set.

The stochastic process is said to be *continuous-parameter*, if the index set is a continuum, and it is said to be *discrete-parameter* if the set is countable.

Inasmuch as in many applications time is the parameter, one then speaks of *continuous-time* or *discrete-time* stochastic processes, the parameter will be considered scalar-valued and denoted by $t \in T$ in what follows.

A stochastic process $\{X(t), t \in T\}$ may be described (*) by specifying the joint probability law of the random variables $X(t_1)$, $X(t_2)$,..., $X(t_n)$ for all integers n and all parameter values $t_1, t_2, \dots t_n \in T$.

This can be done, for instance, by specifying the n-dimensional distribution functions given, for all real $x_1, x_2, \dots, x_n$ by

$$F_{X(t_1)X(t_2)\dots X(t_n)}(u_1, u_2, \dots, u_n) = \tag{C-1}$$
$$= P[X(t_1) \leq x_1, X(t_2) \leq x_2, \dots, X(t_n) \leq x_n],$$

(*) Strictly speaking, the family of finite-dimensional probability laws defines the probability law of $X(t)$ for all Borel sets of $X(t)$.

or equivalently by specifying the n-dimensional joint characteristic functions given, for all real $u_1, u_2, \ldots, u_n$, by

$$\varphi_{X(t_1)X(t_2)\ldots X(t_n)}(u_1, u_2, \ldots, u_n) = \tag{C.2}$$
$$= E\{\exp[i(u_1X(t_1) + u_2X(t_2) + \ldots + u_nX(t_n))]\} .$$

If the random variables $X(t_1)$, $X(t_2)$, ... $X(t_n)$, for all integers n and all $t_1, t_2, \ldots, t_n \in T$ are continuous the stochastic process is said to be <u>continuous-state</u>, if they are discrete the stochastic process is said to be <u>discrete-state</u>.

A continuous-state stochastic process $X(t)$ may also be described by specifying the n-<u>dimensional probability density functions</u> defined for all real $x_1, x_2, \ldots, x_n$ by

$$f_{X(t_1)X(t_2)\ldots X(t_n)}(x_1, x_2, \ldots x_n) = \tag{C.3}$$
$$= \frac{\partial^n F_{X(t_1)X(t_2)\ldots X(t_n)}(x_1, x_2, \ldots, x_n)}{\partial x_1 \partial x_2 \ldots \partial x_n} .$$

In consequent continuation of what has been said for random variables, the stochastic process $X(t)$ may also be described by the <u>moment functions</u>

$$\alpha_1\{X(t_1)\} = E\{X(t_1)\} ,$$
$$\alpha_2\{X(t_1), X(t_2)\} = E\{X(t_1)X(t_2)\} , \tag{C.4}$$
$$\vdots$$

by the first moment function and the <u>central moment functions</u> of higher than first order, i.e. by

$$\alpha_1\{X(t_1)\},$$

$$\text{(C.5)}\quad \mu_2\{X(t_1),X(t_2)\} = E\{[X(t_1)-E\{X(t_1)\}][X(t_2)-E\{X(t_2)\}]\},$$

$$\vdots$$

or by the cumulant functions

$$\text{(C.6)}\quad \varkappa_n\{X(t_1),X(t_2),\ldots,X(t_n)\} = i^{-n}\frac{\partial^n \log \varphi_{X(t_1)X(t_2)\ldots X(t_n)}(u_1,u_2,\ldots,u_n)}{\partial u_1 \partial u_2 \ldots \partial u_n}.$$

provided these functions exist. This can be seen from the series expansions

$$\text{(C.7a)}\quad \varphi_{X(t_1)X(t_2)\ldots X(t_n)}(u_1,u_2,\ldots,u_n) = 1 + iu_j E\{X(t_j)\} + \frac{1}{2!}(iu_j)(iu_k)E\{X(t_j)\,X(t_k)\} + \ldots,$$

or

$$\text{(C.7b)}\quad \log \varphi_{X(t_1)X(t_2)\ldots X(t_n)}(u_1,u_2,\ldots,u_n) = (iu_j)\varkappa_1\ X(t_j)\} + \frac{1}{2!}(iu_j)(iu_k)\varkappa_2\{X(t_j)X(t_k)\} + \ldots,$$

where the summation convention has been used, and where the indices j, k,... range from 1 through n.

Of special importance in practical applications are the first- and second-order moment functions

$$m_X(t) = E\{X(t)\}\,,$$
$$R_{XX}(t_1,t_2) = \alpha_2\{X(t_1),X(t_2)\}\,, \tag{C.8}$$

called <u>mean</u> (<u>value</u>) <u>function</u> and <u>autocorrelation function</u>, respectively, and the second cumulant function called <u>autocovariance function</u>, $C_{XX}(t_1,t_2) = \varkappa_2\{X(t_1),X(t_2)\}$, which can be identified to be identical with the second central moment function:

$$\begin{aligned} C_{XX}(t_1,t_2) &= E\{[X(t_1)-E\{X(t_1)\}][X(t_2)-E\{X(t_2)\}]\} \\ &= E\{X(t_1)X(t_2)\} - m_X(t_1)m_X(t_2) \\ &= R_{XX}(t_1,t_2) - m_X(t_1)m_X(t_2)\,. \end{aligned} \tag{C.9}$$

The <u>variance function</u> is defined by

$$\begin{aligned} \sigma^2\{X(t)\} &= \mathrm{Var}\{X(t)\} = C_{XX}(t,t) \\ &= R_{XX}(t,t) - m_X^2(t)\,. \end{aligned} \tag{C.10}$$

The joint behavior of two stochastic processes $\{X_1(t), t\in T\}$ and $\{X_2(t), t\in T\}$ may be described by specifying the joint probability law of the random variables $X_1(t_1),\dots,X_1(t_n)$ and $X_2(t_1'),\dots,X_2(t_m')$ for all integers n and m , and all parameter values $t_1,\dots,t_n\,,\,t_1',\dots,t_m' \in T$. Again this can be done by specifying the $(n+m)$-dimensional distribution functions, characteristic functions, probability density functions (in the continuous-state case), the moment functions, or the cumulant functions.

The extension to more than two stochastic processes is straightforward.

The counterparts of autocorrelation function and autocovariance function, where $X(t_1)$ and $X(t_2)$ belong to the same stochastic process, are the <u>crosscorrelation function</u>

$$R_{X_1X_2}(t_1,t_2) = E\{X_1(t_1)X_2(t_2)\} \tag{C.11}$$

and the <u>crosscovariance function</u>

$$\begin{aligned} C_{X_1X_2}(t_1,t_2) &= E\{[X_1(t_1)-E\{X_1(t_1)\}][X_2(t_2)-E\{X_2(t_2)\}]\} \\ &= E\{X_1(t_1)X_2(t_2)\} - m_{X_1}(t_1)m_{X_2}(t_2) \\ &= R_{X_1X_2}(t_1,t_2) - m_{X_1}(t_1)m_{X_2}(t_2) \end{aligned} \tag{C.12}$$

where $X_1(t_1)$ and $X_2(t_2)$ now belong to different stochastic processes.

The theory which deals with those properties of stochastic processes that may be described by first and second order moments is called <u>correlation theory</u>.

Two stochastic processes $X_1(t)$ and $X_2(t)$ are said to be <u>independent</u> if, for all n, all $t_1, t_2, \ldots, t_n \in T$, and all $t_1', t_2', \ldots, t_n' \in T$, the two groups $X_1(t_1), X_1(t_2), \ldots, X_1(t_n)$ and $X_2(t_1'), X_2(t_2'), \ldots, X_2(t_n')$ are independent. They are said to be <u>uncorrelated</u> if for all $t_1, t_2 \in T$

$$R_{X_1X_2}(t_1,t_2) = m_{X_1}(t_1)m_{X_2}(t_2) \, ,$$

or equivalently if for all $t_1, t_2 \in T$

$$C_{X_1X_2}(t_1,t_2) = 0 .$$

They are said to be <u>orthogonal</u> if for all $t_1, t_2 \in T$ one has

$$R_{X_1X_2}(t_1,t_2) = 0 .$$

Analogue to the notation used for random variables the special moment functions

$$\alpha_n\{\underbrace{X(t), X(t), \ldots, X(t)}_{n}\}$$

and

$$\alpha_n\{\underbrace{X_1(t), \ldots, X_1(t)}_{r_1}, \underbrace{X_2(t), \ldots, X_2(t)}_{r_2 = n - r_1}\}$$

are, for simplicity, frequently denoted by

$$\alpha_n\{X(t)\} ,$$

and

$$\alpha_{r_1, r_2}\{X_1(t), X_2(t)\} ,$$

respectively, and an analogue short-hand notation is used for central moment functions and cumulant functions, and for the case of more than two stochastic processes.

Let the index set T have the property that the sum of any two elements of T also belongs to T. A stochastic process

$\{X(t), t \in T\}$ is then said to be

i, <u>strictly homogeneous of order k</u>, k being a positive integer if for all choices of k parameter values t_1, ..., $t_k \in T$, and every $\tau \in T$, the random variables $X(t_1)$, $X(t_2)$, ..., $X(t_k)$ obey the same probability law as $X(t_1+\tau), X(t_2+\tau), \ldots, X(t_k+\tau)$ do,

ii, <u>strictly homogeneous</u> or <u>strongly homogeneous</u> if it is strictly homogeneous of order k for all k

iii, <u>weakly homogeneous</u>, <u>homogeneous in the wide sense</u>, <u>covariance homogeneous</u>, or <u>second order homogeneous</u>, if it possesses finite second order moment functions, constant mean function and an autocorrelation function (or autocovariance function) that depends only on $|t_2 - t_1|$; i.e. if there exists a function $R_{XX}(.)$ such that for all $t_1, t_2 \in T$ one has

(C.13) $$R_{XX}(t_1,t_2) = R_{XX}(t_1 - t_2) = R_{XX}(t_2 - t_1) .$$

Of course, all strictly homogeneous processes are weakly homogeneous.

If the parameter t stands for time a homogeneous process is called <u>time-homogeneous</u> or <u>stationary</u>.

By definition, correlation functions are symmetric,

(C.14) $$R_{XX}(t_1,t_2) = R_{XX}(t_2,t_1) ,$$
$$R_{X_1X_2}(t_1,t_2) = R_{X_2X_1}(t_2,t_1) ,$$

and for weakly stationary stochastic processes R_{XX} is an even function of the time difference:

$$R_{XX}(t_1,t_2) = R_{XX}(\tau) = R_{XX}(-\tau), \tag{C.15a}$$

but note that

$$R_{X_1X_2}(t_1,t_2) = R_{X_1X_2}(\tau) = R_{X_2X_1}(-\tau), \tag{C.15b}$$

where $\tau = t_1 - t_2$.

Since the autocorrelation function of weakly stationary stochastic processes is also nonnegative definite (*) it has a (nonnegative) Fourier transform, the (<u>mean</u> <u>square</u>) <u>spectral</u> <u>density</u> <u>function</u> or power spectrum:

$$\begin{aligned} S_{XX}(\omega) &= \frac{1}{2\pi}\int_{-\infty}^{\infty} R_{XX}(\tau)e^{-i\omega\tau}d\tau, \\ R_{XX}(\tau) &= \int_{-\infty}^{\infty} S_{XX}(\omega)e^{i\omega\tau}d\omega. \end{aligned} \tag{C.16}$$

These two relations are known as Wiener-Khintchine theorem (**)

(*) For arbitrary functions $h(t)$

$$R_{XX}(t_i - t_j)h(t_i)h^*(t_j) \geq 0,$$

where i and j range from 1 to any finite integer, $h^*(t)$ denotes the complex conjugate of $h(t)$, and where the summation convention is used.

(**) Strictly speaking, the right-hand side of the second equation coincides with the left-hand side only at points of continuity of $R_{XX}(\tau)$ provided that $R_{XX}(\tau)$ is piecewise continuous with bounded variation.

If the spectral density functions $S_{X_1X_1}(\omega)$ and $S_{X_2X_2}(\omega)$ of two weakly stationary stochastic processes $X_1(t)$ and $X_2(t)$ exist, the <u>cross spectral density function</u> $S_{X_1X_2}(\omega)$, defined for all real ω by

(C.17a) $$S_{X_1X_2}(\omega) = \frac{1}{2\pi}\int_{-\infty}^{\infty} R_{X_1X_2}(\tau)e^{-i\omega\tau}d\tau ,$$

exists, and

(C.17b) $$R_{X_1X_2}(\tau) = \int_{-\infty}^{\infty} S_{X_1X_2}(\omega)e^{i\omega\tau}d\omega$$

follows.

From Schwarz's inequality, there follows that, for weakly stationary stochastic processes $X_1(t)$ and $X_2(t)$,

(C.18) $$|R_{X_1X_2}(\tau)| \leqslant \sqrt{R_{X_1X_1}(0)R_{X_2X_2}(0)}$$

and, replacing $X_1(t)$ and $X_2(t)$ by $X(t)$, one obtains

(C.19) $$|R_{XX}(\tau)| \leqslant R_{XX}(0) .$$

Of course, all these properties of correlation functions are also valid for covariance functions.

A very important class of stochastic processes are <u>Markov processes</u>, because of their similarity to the deterministic processes of classical dynamics, inasmuch as, intuitively speaking, the probability law of the future development of a Markov process given its present state does not depend on its past.

A discrete-state stochastic process $X(t)$ is said to be <u>Markovian</u> (<u>Markov chain</u>) if for all integers n and all times $t_n > t_{n-1} > \ldots > t_2 > t_1 \in T$

$$F_{X(t_n)|X(t_{n-1})\ldots X(t_1)}(x_n|x_{n-1},\ldots,x_1) = \\ = F_{X(t_n)|X(t_{n-1})}(x_n|x_{n-1}) , \tag{C.20}$$

where the <u>conditional</u> <u>distribution</u> <u>function</u> of $X(t)$ given $X(t_1) = x_1, \ldots, X(t_m) = x_m$ is defined by

$$F_{X(t)|X(t_1)\ldots X(t_m)}(x|x_1,\ldots,x_m) = \\ = \frac{F_{X(t)X(t_1)\ldots X(t_m)}(x,x_1,\ldots,x_m)}{F_{X(t_1)\ldots X(t_m)}(x_1,\ldots,x_m)} . \tag{C.21}$$

A continuous-state stochastic process $X(t)$ is said to be Markovian if for all integers n and all times $t_n > t_{n-1} > \ldots > t_2 > t_1 \in T$

$$f_{X(t_n)|X(t_{n-1})\ldots X(t_1)}(x_n|x_{n-1},\ldots,x_1) = \\ = f_{X(t_n)|X(t_{n-1})}(x_n|x_{n-1}) , \tag{C.22}$$

where the <u>conditional</u> <u>density</u> <u>function</u> of $X(t)$ given $X(t_1) = x_1, \ldots, X(t_m) = x_m$ is defined by

$$f_{X(t)|X(t_1)\ldots X(t_m)}(x|x_1,\ldots,x_m) = \\ = \frac{f_{X(t)X(t_1)\ldots X(t_m)}(x,x_1,\ldots,x_m)}{f_{X(t_1)\ldots X(t_m)}(x_1,\ldots,x_m)} . \tag{C.23}$$

The special conditional density function $f_{X(t)|X(t_0)}(x|x_0)$ is called transition (probability) density function.

The conditional expectation of a function $g[X(t)]$, where $X(t)$ is a continuous-state Markov process, given $X(t_0) = x_0$, $t \geq t_0$, is defined by

$$E\{g[X(t)]|X(t_0) = x_0\} = \int g(x) f_{X(t)|X(t_0)}(x|x_0)dx . \tag{C.24}$$

The transition density function satisfies the equation

$$\frac{\partial f_{X(t)|X(t_0)}(x|x_0)}{\partial t} = \sum_{n=1}^{\infty} \frac{(-1)^n}{n!} \frac{\partial^n}{\partial x^n}[K_n(x,t) f_{X(t)|X(t_0)}(x|x_0)] , \tag{C.25}$$

where the derivate moments $K_n(x,t)$ are defined by

$$K_n(x,t) = \lim_{\tau \to 0} \frac{1}{\tau} E\{[X(t+\tau) - X(t)]^n | X(t) = x\} , \tag{C.25a}$$

provided that the limits exist. In the case where the sum in Eq. (C.25) has only a finite number of nonvanishing terms, Eq. (C.25) is a linear partial differential equation; in the case of an infinite number of terms the sum is equivalent to an integral operator.

The derivate moments $K_n(x,t)$ of a continuous Markov process vanish for $n \geq 3$, Eq. (C.25) then reads

$$\frac{\partial f_{X(t)|X(t_0)}(x|x_0)}{\partial t} = -\frac{\partial}{\partial x}\left[K_1(x,t) f_{X(t)|X(t_0)}(x|x_0)\right] + \\ + \frac{1}{2}\frac{\partial^2}{\partial x^2}\left[K_2(x,t) f_{X(t)|X(t_0)}(x|x_0)\right] \tag{C.26}$$

and is called Fokker-Planck equation or diffusion equation.

A stochastic process $\{X(t), t \in T\}$ is said to be normal or Gaussian, if the random variables $X(t_1), X(t_2), \ldots, X(t_n)$ for all integers n and all $t_1, t_2, \ldots, t_n \in T$, are jointly normally distributed. For such a process the whole probability law is determined if the first and second order moment functions are known.

A continuous-parameter stochastic process $\{X(t), t \in T\}$ is said to have independent increments if for all integer n and all parameter values $t_1, t_2, \ldots, t_n \in T$ the random variables

$$X(t_2) - X(t_1), \ldots, X(t_n) - X(t_{n-1})$$

are independent; the process is said to have stationary independent increments if it has independent increments and if for all $t_1, t_2 \in T$, with $t_1 < t_2$, and every $\tau > 0$ for which $t_1 + \tau$, $t_2 + \tau \in T$ the random variables $X(t_2) - X(t_1)$ and $X(t_2 + \tau) - X(t_1 + \tau)$ have the same distribution.

A continuous-parameter stochastic process $\{X(t), t \geq 0\}$ which, in addition of being normal, has

i, stationary independent increments

ii, vanishing mean function, $E\{X(t)\} = 0$ for every $t \geq 0$, and

iii, vanishing initial value, $X(0) = 0$,

is called Wiener process or Brownian motion process. This important process is nonstationary, continuous but, strictly speaking, nowhere differentiable.

Weakly stationary stochastic processes may also be characterized by the behavior of their spectral density functions.

A weakly stationary stochastic process $X(t)$ is said to be band-limited if $S_{XX}(\omega)=0$ for all $|\omega| > \omega_c$, it is said to be low-pass if it is band-limited and $S_{XX}(\omega)$ is substantial over an interval extending from zero to some upper cut-off frequency, and it is said to be ideal low-pass if

$$S_{XX}(\omega) = \begin{cases} S_0 & |\omega| \leq \omega_c \\ 0 & |\omega| > \omega_c \, . \end{cases} \tag{C.27}$$

A weakly stationary stochastic process $X(t)$ is said to be bandpass or bandwidth-limited if

$$S_{XX}(\omega) = 0 \text{ for all } \omega \text{ which } \omega_\ell > |\omega| > \omega_u \, ,$$

where $0 \leq \omega_\ell < \omega_u < \infty$, and it is said to be ideal bandpass if

$$S_{XX}(\omega) = \begin{cases} S_0 & \omega_\ell \leq |\omega| \leq \omega_u \,, \\ 0 & \text{elsewhere} \,. \end{cases} \tag{C.28}$$

A weakly stationary stochastic process is said to be narrow-band if it is bandpass and $\omega_u - \omega_\ell << (\omega_\ell + \omega_u)/2$, and it is said to be (ideal) broad-band if it is (ideal) bandpass and the bandwidth $\omega_u - \omega_\ell$ is at least of the same order of magnitude as the central frequency $(\omega_\ell + \omega_u)/2$.

A weakly stationary process $X(t)$ is said to be white noise if

$$S_{XX}(\omega) = S_0 \qquad -\infty < \omega < \infty \,. \tag{C.29}$$

Such a process is physically not realizable inasmuch as it would possess infinite average "energy":

$$R_{XX}(\tau) = 2\pi S_0 \delta(\tau) \tag{C.30}$$

where $\delta(\tau)$ is Dirac's delta function, but it is a mathematical idealization, as limit case of an ideal broad-band stochastic process, which leads in many practical applications to meaningful results.

A continuous-parameter stochastic process $X(t)$ is said to be mean square continuous at $t \in T$ if

$$\underset{\tau \to 0}{\mathrm{l.i.m.}}\, X(t + \tau) = X(t) \,. \tag{C.31}$$

Necessary and sufficient for mean square continuity of $X(t)$ at $t \in T$ is that the autocorrelation function $R_{XX}(t_1,t_2)$ is finite and continuous on the diagonal, for $t_1 = t_2 = t$.

The <u>mean square derivative</u> of a continuous parameter stochastic process $X(t)$ at $t \in T$ is defined by

$$X'(t) = \underset{\tau \to 0}{\text{l.i.m.}} \frac{X(t+\tau) - X(t)}{\tau} . \tag{C.32}$$

Necessary and sufficient for $X(t)$ to be differentiable at $t \in T$ is that the autocorrelation function $R_{XX}(t_1,t_2)$ has a continuous mixed second derivative on the diagonal, for $t_1 = t_2 = t$.

The operations of taking expectation and l.i.m. commute. There follows that

$$\frac{d}{dt} m_X(t) = m_{X'}(t) , \tag{C.33}$$

$$\frac{\partial}{\partial t_1} R_{XX}(t_1,t_2) = R_{X'X}(t_1,t_2) ,$$

$$\frac{\partial^2}{\partial t_1 \partial t_2} R_{XX}(t_1,t_2) = R_{X'X'}(t_1,t_2) ,$$

and so on. If $X(t)$ is weakly stationary one has

$$R_{X'X'}(t_1,t_2) = -R''_{XX}(t_2-t_1) = R_{X'X'}(t_2-t_1),$$
$$R_{X'X}(t_1,t_2) = R'_{XX}(t_2-t_1) = R_{X'X}(t_2-t_1), \tag{C.34}$$

where a prime associated with R denotes differentiation with respect to the argument. Since $R_{XX}(t_2-t_1)$ is an even function there follows that

$$R_{X'X}(0) = R'_{XX}(0) = 0\ , \tag{C.35}$$

the random variables $X(t)$ and $X'(t)$ are orthogonal, for all differentiable weakly stationary stochastic processes $X(t)$ and for all $t \in T$.

Integrals of the form

$$\int_a^b X(t)h(t,\tau)dt \tag{C.36}$$

where $X(t)$ is a continuous-parameter stochastic process and $h(t,\tau)$ a bounded, possibly complex-valued, deterministic function are called <u>stochastic integrals</u>. These integrals are well defined as mean square limit of Riemannian sums if

$$\int_a^b\int_a^b R_{XX}(t_1,t_2)h(t_1,\tau_1)h(t_2,\tau_2)dt_1dt_2$$

is bounded for all τ_1 and τ_2 , and the linear operations of integration and taking expectation commute then.

For

$$Y(\tau) = \int_a^b X(t)h(t,\tau)dt, \tag{C.37}$$

where $h(t,\tau)$ is real, one has

$$\begin{aligned} &\alpha_n\{Y(\tau_1),\ldots,Y(\tau_n)\} = \\ &= \int_a^b \ldots \int_a^b \alpha_n\{X(t_1),\ldots,X(t_n)\}h(t_1,\tau_1)\ldots h(t_n,\tau_n)dt_1\ldots dt_n \end{aligned} \tag{C.38}$$

and

$$\begin{aligned} &\varkappa_n\{Y(\tau_1),\ldots,Y(\tau_n)\} = \\ &= \int_a^b \ldots \int_a^b \varkappa_n\{X(t_1),\ldots,X(t_n)\}h(t_1,\tau_1)\ldots h(t_n,\tau_n)dt_1\ldots dt_n, \end{aligned} \tag{C.39}$$

for any nonnegative integers n and m, provided that the right-hand sides exist. From these general results the relations

$$\begin{aligned} m_Y(\tau) &= \int_a^b m_X(t)h(t,\tau)dt, \\ R_{YY}(\tau_1,\tau_2) &= \int_a^b\int_a^b R_{XX}(t_1,t_2)h(t_1,\tau_1)h(t_2,\tau_2)dt_1dt_2 \end{aligned} \tag{C.40}$$

follow immediately.

Most of the definitions and theorems given in this paragraph may easily be extended to vector-valued stochastic processes, inasmuch as the components of the latters are (scalar-valued) stochastic processes, and to stochastic fields (more than one indexing parameter, or vector-valued parameters).

Literature

[1] Lin, Y.K.: Probabilistic Theory of Structural Dynamics, New York: McGraw-Hill, 1967

[2] Papoulis, A.: Probability, Random Variables, and Stochastic Processes, New York: McGraw-Hill, 1967

[3] Parzen, E.: Modern Probability Theory and Its Applications, New York: Wiley, 1960

[4] Parzen, E.: Stochastic Processes, 3rd Printing, San Francisco: Holden-Day, 1967

D. Transformation of Stochastic Processes through Linear Systems

Memoryless Systems

Let the input-output relationship of a linear single-input-single-output system be given by

$$Y(t) = a(t)U(t) + b(t)\ , \quad t \geq t_0 \tag{D.1}$$

where $U(t)$ is a continuous-state stochastic process and $a(t)$ and $b(t)$ are deterministic functions.

Equation (B.45) gives immediately

$$\begin{aligned} & f_{Y(t_1)\ldots Y(t_k)}(y_1,\ldots,y_k) = \\ & = [a(t_1)\ldots a(t_k)]^{-1} f_{U(t_1)\ldots U(t_n)}(u_1,\ldots,u_k) \end{aligned} \tag{D.2}$$

for all $t_i \geq t_0$ for which $a(t_i) \neq 0$, and where

$$u_i = (y_i - b(t_i))/a(t_i)\ , \quad i = 1,\ldots,k\ .$$

The moment functions of the output process $Y(t)$ may be computed using the recursion formula

$$\begin{aligned} & \alpha_{k+\ell}\{Y(t_1),\ldots,Y(t_k),U(t_{k+1}),\ldots,U(t_{k+\ell})\} = \\ & = a(t_k)\alpha_{k+\ell}\{Y(t_1),\ldots,Y(t_{k-1}),U(t_k),\ldots,U(t_{k+\ell})\} \\ & + b(t_k)\alpha_{k+\ell-1}\{Y(t_1),\ldots,Y(t_{k-1}),U(t_{k+1}),\ldots,U(t_{k+\ell})\}\ , \end{aligned} \tag{D.3a}$$

or upon usage of

$$\alpha_{k+\ell}\{Y(t_1),\ldots,Y(t_k),U(t_{k+1}),\ldots,U(t_{k+\ell})\} =$$

$$= \prod_{i=1}^{k} a(t_i)\alpha_{k+\ell}\{U(t_1),\ldots,U(t_{k+\ell})\}$$

$$+ b(t_k)\alpha_{k+\ell-1}\{Y(t_1),\ldots,Y(t_{k-1}),U(t_{k+1}),\ldots,U(t_{k+\ell})\} \tag{D.3b}$$

$$+ b(t_1)\prod_{i=0}^{k-2} a(t_{k-i})\alpha_{k+\ell-1}\{U(t_2),\ldots,U(t_{k+\ell})\}$$

$$+ \sum_{i=1}^{k-2} b(t_{k-i})\prod_{j=0}^{i-1} a(t_{k-j})\alpha_{k+\ell-1}\{Y(t_1),\ldots,Y(t_{k-i-1}),U(t_{k-i+1}),\ldots,U(t_{k+\ell})\} .$$

Both recursion formulae are valid for all $k \geq 1$, all $\ell \geq 0$, and all $t_i \geq t_0$, provided the right-hand sides exist.

For $k = 0$, $\ell = 0$ and $k = 2$, $\ell = 0$ there follows

$$m_Y(t) = a(t)m_U(t) + b(t) ,$$

$$R_{YY}(t_1,t_2) = a(t_1)a(t_2)R_{UU}(t_1,t_2) + a(t_1)b(t_2)m_U(t_1) \tag{D.4}$$

$$+ a(t_2)b(t_1)m_U(t_2) + b(t_1)b(t_2) ,$$

and for the special moment functions $\alpha_k\{Y(t)\}$ one has

$$\alpha_k\{Y(t)\} = \sum_{i=0}^{k} \binom{k}{i} b(t)^i a(t)^{k-i} \alpha_{k-i}\{U(t)\} . \tag{D.5}$$

Differential Systems:

Let a single-input differential system be characterized by

(D.6) $$N(D,t)Y(t) = M(D,t)U(t),$$

where $U(t)$ is a continuous-state continuous-time stochastic process, and where the linear deterministic operators $N(D,t), M(D,t)$ are of the form

(D.7) $$N(D,t) = \sum_{i=0}^{n} a_i(t)D^i, \quad M(D,t) = \sum_{i=0}^{m} b_i(t)D^i,$$

with D^i standing for d^i/dt^i.

Differential equations of the moment functions:

Taking expectation of Eq. (D.6) and interchanging the order of taking expectation and differentiation one obtains

(D.8) $$N(D,t)m_Y(t) = M(D,t)m_U(t).$$

Replacement of t by t_i and $D = d/dt$ by $D_i = d/dt_i$ in Eq. (D.6) renders

$$N(D_i,t_i)Y(t_i) = M(D_i,t_i)U(t_i).$$

Multiplying now k left- and k right-hand sides, for $i = 1,2,\dots,k$, respectively, the result by $U(t_{k+1})\dots U(t_{k+\ell})$ and taking expectation one is lead to the relation

$$N(D_1,t_1)\ldots N(D_k,t_k)\alpha_{k+\ell}\{Y(t_1),\ldots,Y(t_k),U(t_{k+1}),\ldots,U(t_{k+\ell})\} =$$

$$= M(D_1,t_1)\ldots M(D_k,t_k)\alpha_{k+\ell}\{U(t_1),\ldots,U(t_{k+\ell})\}\,, \tag{D.9}$$

provided that $U(t)$ is m times differentiable and the moment on the right-hand side exists. This equation is valid for all t_i for which (D.6) holds.

For $k=2$, $\ell=0$ and $k=1$, $\ell=1$

$$\begin{aligned} N(D_1,t_1)N(D_2,t_2)R_{YY}(t_1,t_2) &= \\ = M(D_1,t_1)M(D_2,t_2)R_{UU}(t_1,t_2) \end{aligned} \tag{D.10}$$

and

$$N(D_1,t_1)R_{YU}(t_1,t_2) = M(D_1,t_1)R(t_1,t_2) \tag{D.11}$$

follows.

Moment and cumulant functions of the output process:

According to the fundamental decomposition theorem the solution of Eq. (D.6) may be represented as sum of a zero-input term, depending on the initial conditions, and the zero-state response. For deterministic initial conditions the first term is deterministic. It may be treated separately and superposed, if necessary, using the results given for the system characterized by Eq. (D.1). For convenience, we therefore consider the zero-state response of the system only.

Let $h(t,\tau)$ be the impulse-response function of the system. The zero-state response may then be represented in the form

$$Y_0(t) = \int_{t_0}^{t} h(t,\lambda)U(\lambda)d\lambda\ , \quad t \geq t_0\ , \tag{D.12}$$

see (A.9). Equations (C. 36) and (C.37), valid for moment and cumulant functions of stochastic integrals of this type, give for $k \geq 1$, $\ell \geq 0$ and all $t_i \geq t_0$ the relations

$$\begin{aligned} &\alpha_{k+\ell}\{Y_0(t_1),\dots,Y_0(t_k),U(t_{k+1}),\dots,U(t_{k+\ell})\} = \\ &= \int_{t_0}^{t_1}\dots\int_{t_0}^{t_k}\alpha_{k+\ell}\{U(t_1),\dots,U(t_{k+\ell})\}\cdot \\ &\qquad h(t_1,\lambda_1)\dots h(t_k,\lambda_k)d\lambda_1\dots d\lambda_k\ , \end{aligned} \tag{D.13}$$

and

$$\begin{aligned} &\varkappa_{k+\ell}\{Y_0(t_1),\dots,Y_0(t_k),U(t_{k+1}),\dots,U(t_{k+\ell})\} = \\ &= \int_{t_0}^{t_1}\dots\int_{t_0}^{t_k}\varkappa_{k+\ell}\{U(t_1),\dots,U(t_{k+\ell})\}\cdot \\ &\qquad h(t_1,\lambda_1)\dots h(t_k,\lambda_k)d\lambda_1\dots d\lambda_k\ , \end{aligned} \tag{D.14}$$

provided that the moment functions of the integrand and the integrals exist.

For $k=1$, $\ell=0$ and $k=2$, $\ell=0$ there follows

$$m_{Y_0}(t) = \int_{t_0}^{t} m_U(\lambda)h(t,\lambda)d\lambda\,, \tag{D.15a}$$

$$R_{Y_0Y_0}(t_1,t_2) = \int_{t_0}^{t_1}\int_{t_0}^{t_2} R_{UU}(\lambda_1,\lambda_2)h(t_1,\lambda_1)h(t_2,\lambda_1)d\lambda_2 d\lambda_1\,, \tag{D.15b}$$

and

$$C_{Y_0Y_0}(t_1,t_2) = \int_{t_0}^{t_1}\int_{t_0}^{t_2} C_{UU}(\lambda_1,\lambda_2)h(t_1,\lambda_1)h(t_2,\lambda_2)d\lambda_2 d\lambda_1\,. \tag{D.15c}$$

For $m=1$, $n=1$ the crosscorrelation function and the crosscovariance function, respectively, follow as

$$R_{Y_0U}(t_1,t_2) = \int_{t_0}^{t_1} R_{UU}(\lambda_1,t_2)h(t_1,\lambda_1)d\lambda_1\,, \tag{D.16a}$$

$$C_{Y_0U}(t_1,t_2) = \int_{t_0}^{t_1} C_{UU}(\lambda_1,t_2)h(t_1,\lambda_1)d\lambda_1\,. \tag{D.16b}$$

The stochastic integral (D.12) corresponds to the stochastic differential equation (D.6), which involves derivatives. Nevertheless, the stochastic integral may exist even if derivatives of $U(t)$, appearing in (D.6), do not exist.

Stationarity of the output process:

The process $Y_0(t)$, given by Eq. (D.12), is in general nonstationary, but if the system is time-invariant, if the impulse response function, which is a function of $t-\lambda$ then, is absolutely integrable over the interval $[-\infty,t]$,

$$\int_{-\infty}^{t} |h(t-\lambda)| d\lambda < \infty ,$$

and if $U(t)$ is strictly (weakly) stationary $Y_0(t)$ is asymptotically strictly (weakly) stationary, the steady-state response being a strictly (weakly) stationary stochastic process.

In this case there follows, for the mean function of the steady-state response $Y_\infty(t)$, from Eq. (D.15a),

$$m_{Y_\infty}(t) \equiv m_{Y_\infty} = m_U \int_{-\infty}^{\infty} h(\tau) d\tau = m_U H(0) ,$$

where $H(.)$ is the transfer function of the system, and where $h(t) = 0$ for $t < 0$ and $m_U(t) \equiv m_U$ has been used.

Spectral densities of the steady-state response:

In the above mentioned case, where the steady-state response of the system is a stationary stochastic process, the correlation function of $Y_\infty(t)$ may also be obtained via the spectral density functions.

With $t_1 - t_2 = \tau$ and the change of variables $t_1 - \lambda_1 = \lambda$, Eq. (D.16a) may in this case be rewritten in the form

$$R_{Y_\infty U}(\tau) = \int_{-\infty}^{\infty} h(\lambda) R_{UU}(\tau - \lambda) d\lambda .$$

Taking Fourier transform of this equation one obtains

$$S_{Y_\infty U}(\omega) = \frac{1}{2\pi} \int_{-\infty}^{\infty} \int_{-\infty}^{\infty} h(\lambda) R_{UU}(\tau - \lambda) e^{-i\omega\tau} d\lambda d\tau =$$

$$= \frac{1}{2\pi}\int_{-\infty}^{\infty} h(\lambda)e^{-i\omega\lambda}d\lambda \int_{-\infty}^{\infty} e^{-i\omega(\tau-\lambda)}R_{UU}(\tau-\lambda)d\tau$$

and thus finally, with Eqs.(C.17a) and (A.10),

$$S_{Y_\infty U}(\omega) = H(i\omega)S_{UU}(\omega)\,, \tag{D.17}$$

where again $H(.)$ is the transfer function of the system.

From Eqs. (D.15b) and (D.16a) there follows

$$R_{Y_0Y_0}(t_1,t_2) = \int_{t_0}^{t_2} R_{Y_0U}(t_1,\lambda_2)h(t_2,\lambda_2)d\lambda_2\,.$$

Transforming as above one obtains

$$S_{Y_\infty Y_\infty}(\omega) = H(-i\omega)S_{Y_\infty U}(\omega)\,. \tag{D.18}$$

Combination of the two results gives

$$S_{Y_\infty Y_\infty}(\omega) = H(i\omega)H(-i\omega)S_{UU}(\omega) = |H(i\omega)|^2 S_{UU}(\omega)\,, \tag{D.19}$$

the fundamental theorem for the spectral density of the steady-state response of a time-invariant differential system, the impulse-response function of which is absolutely integrable. Analogue to what has been said about the stochastic integral (D.12), this fundamental theorem is valid even in cases where derivatives of $U(t)$ required in (D.6) do not exist. This may be shown [2] , pp. 122, by introducing the column vector $\underline{X}(t)$, the n components of which are stochastic processes $X_1(t),\dots,X_n(t)$. Let n-vectors $\underline{b}$ and $\underline{c}$, and the $n\times n$ matrix $\underline{A}$ be defined by

$$\underline{b}^T = [0,\ldots,0,\beta],$$

$$\underline{c}^T = [b_0 - \gamma a_0, \ldots, b_{n-1} - \gamma a_{n-1}],$$

$$\underline{A} = \begin{vmatrix} 0 & 1 & 0 & . & . & . & 0 \\ 0 & 0 & 1 & . & . & . & . \\ . & . & . & . & . & . & . \\ . & . & . & . & . & . & . \\ . & . & . & . & . & . & 0 \\ 0 & . & . & . & . & 0 & 1 \\ -\alpha_0 & -\alpha_1 & . & . & . & . & -\alpha_{n-1} \end{vmatrix}$$

where a_i and b_i are the coefficients (*) in $N(D)$ and $M(D)$, see (D.7), and $\alpha_i = a_i/a_n$, $\beta = 1/a_n$, and $\gamma = b_n/a_n$. The system of canonical state equations

$$\dot{\underline{X}}(t) = \underline{A}\underline{X}(t) + \underline{b}U(t)$$

$$Y(t) = \underline{c}^T\underline{X}(t) + \alpha_n U(t)$$

is completely equivalent to (D.6) but now there are no derivatives of $U(t)$ involved. A completely different proof for differential systems characterized by (D.6) for which $m < n-1$ may be

(*) For $m<n$ the constants $b_i=0$ for $i = m+1,\ldots,n$.

found in [1], [3].

<u>Markov vector approach</u>

Let the state equations of a linear single-input system be given by

$$\underline{X}(t) = \underline{A}(t)\underline{X}(t) + \underline{b}(t)W(t) , \qquad (D.20)$$

where the state vector $\underline{X}(t)$ and the deterministic vector $\underline{b}(t)$ have n components, $\underline{A}(t)$ is a deterministic $n \times n$ matrix and $W(t)$ is Gaussian white noise with zero mean and spectral density 2D. The mean functions of the components of the zero-state response $\underline{X}_0(t)$ vanish then.

Let the state transition matrix again be denoted by $\underline{\Phi}(t_1,t_2)$, such that

$$\underline{X}_0(t) = \int_{t_0}^{t} \underline{\Phi}(t,\tau)\underline{b}(\tau)W(\tau)d\tau , \quad t \geq t_0 > t_c .$$

The matrix $\underline{R}(t_1,t_2)$ of the correlation functions of the components of $\underline{X}_0(t)$,

$$\underline{R}(t_1,t_2) = E\{\underline{X}_0(t_1)\underline{X}_0^T(t_2)\} \qquad (D.21)$$

is, for $t_2 \geq t_1 > t_0$, then given by

$$\underline{R}(t_1,t_2) = 2D\int_{t_0}^{t_1} \underline{\Phi}(t_1,\tau)\underline{b}(\tau)\underline{b}^T(\tau)\underline{\Phi}^T(t_2,\tau)d\tau . \qquad (D.22)$$

The special result

$$\underline{R}(t,t) = 2D \int_{t_0}^{t} \underline{\Phi}(t,\tau)\underline{b}(\tau)\underline{b}^T(\tau)\underline{\Phi}^T(t,\tau)d\tau \ , \tag{D.23}$$

for $t_2=t_1=t>t_0$, may also be obtained via the Fokker-Planck equation:

Let, in the general (nonlinear) case, differential state equations be given by

$$\dot{X}_i = a_i(\underline{X},t) + b_{ik}(\underline{X},t)W_k(t) \ , \quad i = 1,\dots,n \ , \tag{D.24}$$

where the summation convention has been used. The functions a_i and b_{ik} are assumed to be continuous functions of the n components $X_i(t)$ of the state vector $\underline{X}(t)$ and $W_k(t)$, $k=1,\dots,N$, are Gaussian white noise processes with

$$E\{W_k(t)\} = 0 \ , \quad k = 1,\dots,N \tag{D.25}$$

$$E\{W_k(t_1)W_\ell(t_2)\} = 2D_{k\ell}\,\delta(t_2-t_1) \ , \quad k,\ell = 1,\dots,N \ .$$

The differential equation of the transition density function, the Fokker-Planck equation, reads then (*)

$$\frac{\partial q(\underline{x},t)}{\partial t} = -\frac{\partial}{\partial x_i}\left[a_i(\underline{x},t)q(\underline{x},t)\right] + \tag{D.26}$$

(*) Since white noise is considered as limiting case of a broad band process the so-called physical approach [4] is used .

$$+ D_{k\ell}\frac{\partial}{\partial x_i}\left[b_{ik}(\underline{x},t)\frac{\partial}{\partial x_j}\left[b_{i\ell}(\underline{x},t)q(\underline{x},t)\right]\right], \qquad \text{(D.26)}$$

where $q(\underline{x},t)$ stands for $f_{X(t)|X(t_0)}(x|x_0)$, where the summation convention has been used, and where the indices i,j range from 1 through n , and k,ℓ from 1 trough N, respectively.

In the case under consideration, where the state equations are given by Eq. (D.20), the Fokker-Planck equation (D.26) reduces to

$$\frac{\partial q(\underline{x},t)}{\partial t} = -\frac{\partial}{\partial x_i}\left[A_{ij}(t)x_j q(\underline{x},t)\right] + Db_i(t)b_j(t)\frac{\partial^2 q(\underline{x},t)}{\partial x_i \partial x_j}, \qquad \text{(D.27)}$$

where again the summation convention is used.

Denoting, for simplicity, the conditional moments

$$E\{X_1^{r_1}(t)\ldots X_n^{r_n}(t)|X_1(t_0) = x_{1,0},\ldots,X_n(t_0) = x_{n,0}\}$$

by $\alpha_{r_1,\ldots,r_n}$, one has, using the definition of the conditional expectation, Eq. (C.24),

$$\dot{\alpha}_{r_1,\ldots,r_n} = \int_{-\infty}^{\infty}\int_{-\infty}^{\infty} x_1^{r_1}\ldots x_n^{r_n}\frac{\partial q(\underline{x},t)}{\partial t}\,dx_1\ldots dx_n \,.$$

On replacing $\partial q/\partial t$ with the right-hand side of Eq. (D.27) and integrating by parts one obtains the linear ordinary differential equation

$$\dot{\alpha}_{r_1,\dots,r_n} = \sum_{i=1}^{n} r_i \{A_{ii}\alpha_{r_1,\dots,r_n} + D(r_i-1)b_i^2\alpha_{r_1,\dots,r_i-2,\dots,r_n}\} + \sum_{i,j=1}^{n}{}^{i\neq j} r_i\{A_{ij}\alpha_{r_1,\dots,r_i-1,\dots,r_j+1,\dots,r_n} + Dr_j b_i b_j \alpha_{r_1,\dots,r_i-1,\dots,r_j-1,\dots,r_n}\} \tag{D.28}$$

for the conditional moments of the components of the state vector of a system characterized by Eq. (D.20).

Combining all second order moments to the matrix $\underline{R}(t,t)$, defined by Eq. (D.21), and using the fact that $\underline{R}$ is symmetric, one obtains

$$d\underline{R}(t,t)/dt = \underline{A}(t)\underline{R}(t,t) + \underline{R}(t,t)\underline{A}^T(t) + 2D\underline{b}(t)\underline{b}^T(t) . \tag{D.29}$$

The solution of this equation with initial condition $\underline{R}(t_0,t_0) = \underline{0}$ is given by (D.23).

Concluding remarks:

From the theorems given in this section there follows that the first and second order moment functions of the output of a linear system may be calculated if the first and second order moment functions of the input process and, in case of a differential system, the initial state are known; knowledge of higher order moment functions of the input process is not required. This is especially important in case of Gaussian inputs insofar as the output of a linear system whose input is Gaussian is also Gaussian.

To calculate the mean function of the output of a linear differential system with stochastic input one will, in general, use the very same methods one would use in the case of deterministic inputs.

To calculate the correlation function of the output process the way via spectral densities may be used if only the steady-state response is of interest. If, for the inversion of the Fourier transform, correspondence tables can be used immediately this way is an easy one. If this is not the case the way via the integral representation (D.12) is usually the more efficient one. Using this way one may obtain the correlation function of the nonstationary response with little additional effort. If only the values of the correlation function for $t_2 = t_1 = t$ of the steady-state or of the nonstationary response or higher order moment functions of the form $\alpha_n\{Y(t)\}$ are of interest the Markov vector approach, via Fokker-Planck equation and moment differential equations, is, if applicable, a useful and efficient alternative. The differential equations of moment functions, (D.9) through (D.11), are useful means to check results. In calculating higher order moment functions, of the out put process in case of non-Gaussian inputs, which is a tedious and often questionable task anyway, they are, among the methods mentioned heren the only alternative to the way via the integral representation.

Literature

[1] Lin, Y.K.: Probabilistic Theory of Structural Dynamics, New York: McGraw-Hill, 1967

[2] Papoulis, A.: Probability, Random Variables, and Stochastic Processes, New York: McGraw-Hill, 1965

[3] Parkus, H.: Random Processes in Mechanical Sciences, Udine: CISM, 1969

[4] Stratonovich, R.L.: Topics in the Theory of Random Noise, 2 Vols., New York, Gordon and Breach, 1963

SUBJECT INDEX TO APPENDIX

CONTENTS

Printed in the United States
By Bookmasters